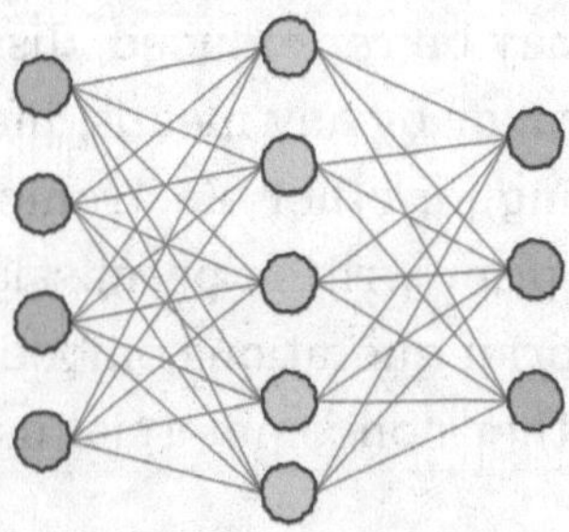

NAVIGATING
Data and AI

A Leader's Guide to Digital Transformation

SARAH CHOUDHARY

Second Edition

Navigating Data and AI, Second Edition

A Leader's Guide to Digital Transformation

This book offers the author's experience and opinions. It is not legal, tax, security, or investment advice. Consult appropriate professionals for your own situation. Technology referenced in this book changes quickly; the principles will outlast the specific tools.

Cover and interior illustrations by the author and her studio.

Set in Georgia.

First edition published 2024.

Second edition published 2026.

ISBN 979-8-9962662-6-5 (Paperback)

ISBN 979-8-9962662-7-2 (Hardcover)

Published by Sarah Choudhary.

For my mother,

who let a curious child take apart the refrigerator,

and never told me to put down the screwdriver.

CHAPTER

Author's Note

I grew up in Islamabad, taking things apart.

The first machine I dismantled was the family refrigerator. I was small, the screwdriver was bigger than my hand, and I had no real plan beyond seeing how the cold air got out. I remember the satisfaction of getting every screw back in place before my mother noticed. The television came later, and I was less lucky with that one.

Two things came out of those early experiments. The first was a lifelong assumption that anything humans build, humans can understand. The second was a piece of luck, which was that nobody around me, especially my mother, ever told me girls shouldn't be doing this. That kind of permission, given without speaking, has shaped most of what I have built since.

Programming was the next layer. In my undergraduate years in Pakistan, I discovered that you could build invisible machines, machines made entirely of language, and they were just as real as the refrigerator had been. My first applied project was a small library cataloguing system, written for a local library. It worked. People used it. I was hooked.

I moved to the United States for graduate school. Masters and PhD in data science. By then, machine learning had stopped being a niche research interest and was starting to become the thing that decided whether you got a credit card, a job interview, a hospital admission, an immigration outcome. I had questions about the responsibility of building systems with that kind of consequence, and graduate school did not have all the answers. Two decades of practice have not given me all the answers either. The questions have only sharpened.

> "Technology is a tool. Tools are neutral about who they harm. The people who build them are not. We do not get to opt out of that responsibility by saying we were just building the tool."
>
> Sarah

This book is for the leader who has decided they need to understand data and AI well enough to make good decisions about it, but who does not have the time or the background to become a practitioner. CIOs. CTOs. VPs of every flavour. Board members. Founders whose products now depend on a layer of machine learning they do not personally maintain. The growing tribe of people who must lead this work without doing it themselves.

My promise to you is the same promise I made in my other two books. I will be specific. I will be honest. I will tell you when something is overrated. I will give you a real glossary, real worksheets, and real examples drawn from work I have done and seen, with the company names removed because the lessons matter and the names usually do not.

A note on the second edition. The first edition of this book was a draft. It said most of the right things, but in the wrong voice, and it leaned too heavily on fictional case studies with company names like SafeBank and MedSecure. I have rewritten it from scratch. The history is unchanged because it is correct. Everything else is new.

So pull up a chair. Open the laptop. Or close the laptop and read this on a printed page. Either way, let's begin.

Sarah Choudhary

Technologist, founder, recovering perfectionist

CHAPTER

Contents

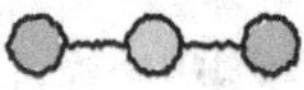

APPENDICES

PART I

The Ground

Where AI came from, what it actually is,
and where it lives.

CHAPTER 1

How We Got Here

A SHORT HISTORY OF AI

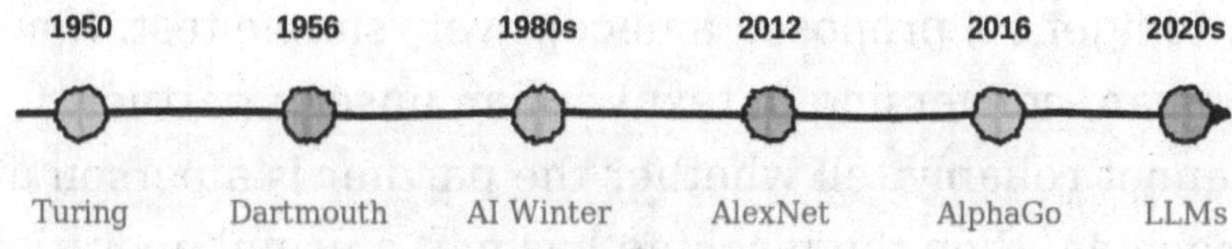

Every conversation about artificial intelligence eventually gets to the question, where did this come from? The short answer is, much earlier than the people in the conversation usually realise.

AI did not arrive in 2022. It arrived, conceptually, in 1950, when a British mathematician named Alan Turing published a paper asking whether machines could think. The technical bones of what we now call machine learning were sketched in the 1950s. The first usable neural network ideas date to the 1960s. By the time the public discovered conversational AI, the field had been at it, on and off, for seventy years.

Understanding that history is not academic. If you are leading an organisation through an AI transition, it matters that the field has been through cycles of hype and disappointment before. We are in a hype phase. We will be in a disappointment phase. Then probably another hype phase. Knowing the rhythm makes you a steadier hand.

Turing's question, and why it still matters

Alan Turing's 1950 paper, Computing Machinery and Intelligence, proposed a deceptively simple test. If a human, conversing in text with an unseen partner, cannot reliably tell whether the partner is a person or a machine, then the machine has met a useful operational definition of intelligence.

Turing was not trying to solve philosophy. He was trying to sidestep it. The question "can machines think" was, in 1950, hopelessly tangled in disputes about consciousness and souls. The Turing Test moved the goalposts to something testable: can a machine behave indistinguishably from a thinking thing? If yes, it is functionally intelligent, regardless of what is happening inside.

Modern large language models can pass a casual Turing Test. This used to be science fiction. The fact that we now barely notice should be your first datum about how fast the ground moves.

Dartmouth, 1956: the field gets a name

In the summer of 1956, a small group of researchers, John McCarthy, Marvin Minsky, Claude Shannon, Nathaniel Rochester, met at Dartmouth College for an eight-week workshop. McCarthy coined the term artificial intelligence in the proposal. The workshop produced no breakthrough algorithms, but it produced a coherent identity for a new field, and the founders of that field went on to shape every major AI laboratory in the United States for the next thirty years.

The opening sentence of the Dartmouth proposal is worth reading slowly. "Every aspect of learning or any other feature of intelligence can in principle be so precisely described that a machine can be made to simulate it." That is either the most exciting or the most chilling sentence in the entire history of computing, depending on how you feel about the project. The field of AI has lived inside that sentence ever since.

The first AI winter, and why it happened

Early enthusiasm overpromised. By the late 1960s, AI researchers had told funders that a machine capable of human-level reasoning was just a decade away. It was not. Funding froze in the 1970s and again in the late 1980s. These periods got a name: AI winter.

The pattern matters because we are watching the seeds of the next one. Every wave of AI enthusiasm has overshot, then under-delivered, then quietly delivered something more modest but more useful, several years late. ELIZA in 1966 was not the conversational AI it appeared to be. Expert systems in the 1980s were not the path to artificial reasoning. Deep learning in the 2010s was not the path to artificial general intelligence. Each of these technologies was real, useful, and oversold. The same is true of today's large language models. Plan accordingly.

> “AI history rhymes. Every wave promises more than it can deliver, then quietly delivers more than its critics expected. Both halves matter.”

Why deep learning broke through when it did

The 2012 ImageNet competition is the moment most people in the field point to as the start of the modern era. A neural network called AlexNet, developed by Alex Krizhevsky, Ilya Sutskever, and Geoffrey Hinton, won the competition by a margin that embarrassed every other approach in the field. Image recognition error rates dropped overnight.

The breakthrough was not really an algorithmic one. Neural networks had existed since the 1960s. Backpropagation, the training technique AlexNet relied on, was published in 1986. What changed was scale.

Three things converged: enough data to train large models (the internet had been quietly producing this for fifteen years), enough computing power to do it cheaply (graphics processing units, originally built for video games, turned out to be exactly the right hardware for matrix multiplication at scale), and enough engineering practice to build the pipelines.

This is the most important practical lesson in the history of AI. Most of the breakthroughs were not the day someone had a new idea. They were the day the world finally had enough compute and enough data to test an idea someone had had decades earlier. If you are wondering when the next breakthrough is coming, watch for the next jump in compute and the next jump in available data. Not the next paper.

AlphaGo, 2016: a turning point most people missed

In March 2016, a system built by DeepMind, then a subsidiary of Google, defeated Lee Sedol, one of the strongest Go players in the world, four games to one. Go had been considered the great unsolved problem in computer games. Its search space is too large to brute-force. The pre-2016 conventional wisdom held that Go-playing software at the human professional level was at least a decade away.

AlphaGo's victory was important less for what it proved about Go and more for what it revealed about reinforcement learning combined with self-play. The system trained partly by playing against itself, millions of games, and developed strategic patterns that even its programmers could not always explain. Move 37 in game two of the Lee Sedol match is now legendary in the Go community: a stone played in a position no human professional would have considered, that turned out to be brilliant. The system saw something humans had not.

If you are looking for the moment when serious technologists started worrying privately about superintelligence, it was around then. Whether they were right to worry is a separate question.

Where we are today, and what to remember

From 2017 onward, the dominant paradigm has been the transformer architecture, which underlies GPT-style models, image generators, code-writing assistants, and most of what the public means when it says "AI." These systems are extraordinarily capable at pattern completion across language, code, image, and structured data. They are not, by any rigorous definition, thinking. They are, however, useful enough that the distinction has stopped mattering to most users.

Four things to take from this chapter:

1. AI is older than its current hype suggests. The cycles of optimism and disappointment have a long pattern.
2. Breakthroughs usually come from compute and data, not new algorithms. Old ideas with new resources beat new ideas with old resources, almost always.

3. Useful does not require thinking. Modern AI systems are extremely useful pattern-matchers. Treating them as more than that leads to bad decisions. Treating them as less than that leads to missed opportunities.
4. Responsibility scales with capability. The more capable a system, the more it matters who built it, on what data, for what purpose, and with what oversight.

A composite case

A logistics company committed to a major AI transformation in early 2017, during the peak of the previous wave of AI enthusiasm. The CEO had read a McKinsey report. The board was excited. They hired an external consulting firm, who recommended a comprehensive program. They spent eighteen million dollars over two years. Most of the projects produced little. By 2019 the team was demoralised, the CEO was on the defensive with the board, and the word "AI" had become an internal punchline. They had hit the disappointment phase of a hype cycle and concluded the technology did not work.

In 2024 a new CEO restarted the AI program. She did it differently. She read the history of the field before she budgeted. She asked her team to identify the specific things that had failed in 2017 and to compare them honestly against what the technology could now do. She ran small experiments before committing to large programs. Her budget for the first year was a tenth of the original, and the focus was three narrow use cases rather than twenty broad ones.

Three years in, the second program has produced measurable returns. The first one, she told me, had failed not because the technology was wrong, but because the leadership had been operating without context. They had not understood that they were arriving at the peak of a hype cycle and would have to weather a disappointment phase before the technology became truly useful. With that context, the second program was designed to survive the disappointment phases that would inevitably come.

Objections you will hear

"We do not need to know the history, we need to focus on the future"

The past is the only data you have. Patterns repeat. Companies that invest at the peak of a hype cycle without understanding they are at the peak tend to make the same mistakes the previous generation made. Leaders who can name the previous cycles tend to budget more conservatively, ship more carefully, and survive the inevitable downturns.

"AI is different this time, history does not apply"

Every wave's leaders have said this. Every wave's leaders have been partly right and substantially wrong. The technology is different. The dynamics of overpromise and underdelivery, the funding cycles, the public-then-disappointed-then-mature pattern, those are durable.

Assume the dynamics will repeat unless you can name a specific reason they will not.

"Our competitors are moving fast, we do not have time for context"

Many of your competitors are about to spend significant money repeating the mistakes of 2017. Moving slightly slower with better context is often the winning move. The companies that ship the most AI in 2026 are not necessarily the companies that benefit most from AI in 2030.

"The history is academic, what matters is the current state"

The current state is one frame in a longer film. Watching only the current frame, you cannot tell whether the action on screen is the beginning, middle, or end of the scene. The history tells you which. Most leaders making AI decisions in 2026 think they are in 2012 (early, untapped opportunity). The honest read is that they are closer to 1999 (genuine transition, with significant froth, and a coming correction that will separate the durable from the disposable).

Questions for your leadership team

5. Can we name, specifically, the previous AI hype cycles? What happened to companies that committed heavily at the peak of each?
6. Where in the current cycle do we believe we are? What evidence supports that view, and what evidence would change it?

7. Which past AI claims at our own company have not panned out? What did we learn that we are applying now?
8. If we are at the peak of this cycle rather than the start, what specifically would we do differently?
9. What is our internal mechanism for distinguishing durable trend from passing hype?

IF YOU REMEMBER NOTHING ELSE FROM THIS CHAPTER

The field has been through cycles of hype and disappointment before. We are in a hype cycle. Plan for the disappointment cycle.

When you wonder if the next big AI breakthrough is coming, look at compute trends and data availability, not at the latest paper.

The systems we have today are powerful pattern matchers, not minds. Build accordingly.

ONE-LINE TAKEAWAY

AI did not start in 2022. Its history is a series of overpromised waves that quietly delivered something useful, several years later than expected. Knowing the pattern makes you a steadier leader during the next one.

CHAPTER 2

The Basics, Plainly

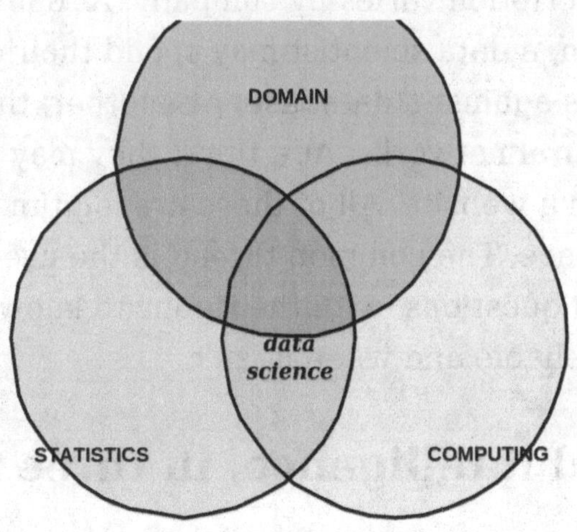

This chapter is the vocabulary you need so the rest of the book makes sense. If you already know what supervised learning and a feature are, skim. If you don't, read carefully. The terms in this chapter come up in every conversation about AI for the rest of your career, and the people using them often don't agree on what they mean.

What data science actually is

Data science sits at the intersection of three things: statistics, computing, and domain knowledge. None of the three on its own is data science. A statistician without a computer is doing classical statistics. A programmer without domain knowledge is building unused tools. A domain expert without quantitative skills is making good guesses. Data science is the practice of combining all three to extract reliable knowledge from data.

The job description varies by company. At one organisation, a data scientist may spend their day writing SQL queries against a database. At another, they may be training neural networks. At a third, they may be running A/B tests on a website. All of these are legitimate forms of data science. The common thread is the use of data to answer real questions, with the rigour to know when the answer is reliable and when it isn't.

Artificial intelligence, in three nested boxes

There are three terms that get used interchangeably and shouldn't be. Here is how they actually nest.

Artificial intelligence (the outer box)

Any technique that gets a machine to perform a task that, if a human did it, we would call intelligent. This includes rule-based expert systems from the 1980s, classical search algorithms, and modern neural networks. It is a very wide tent.

Machine learning (the middle box)

A specific subset of AI in which the machine learns from data rather than being explicitly programmed with rules. Instead of a programmer writing "if income > $50,000 and age < 35, approve loan," a machine learning system is shown thousands of past loan decisions and infers its own decision rules from the patterns.

Deep learning (the inner box)

A specific kind of machine learning that uses neural networks with many layers. It dominates current research and product development because it handles unstructured data, images, audio, language, far better than earlier approaches. Almost everything that is making news in 2026 is deep learning.

Important: all deep learning is machine learning. All machine learning is AI. But not all AI is machine learning, and not all machine learning is deep learning. Get this nesting straight and a lot of confusing vendor pitches become clearer.

Supervised, unsupervised, reinforcement: three ways to learn

Supervised learning

The machine is shown many examples with the right answer attached. Show it ten thousand photos labelled "cat" or "dog," and it learns to classify new photos. The vast majority of practical machine learning is supervised. The bottleneck is usually getting the labelled data.

Unsupervised learning

The machine is shown data without labels and asked to find structure. Cluster these customers into groups that buy similarly. Detect transactions that look unusual. Useful for exploration, less useful for definitive answers.

Reinforcement learning

The machine takes actions in an environment and gets rewarded or penalised. It learns through trial and error to maximise long-term reward. This is how AlphaGo learned. Powerful in narrow, well-defined environments. Hard to use safely in messy real-world ones.

The vocabulary leaders should know cold

A short list. Learn these. The rest of the book uses them freely.

- Model: the mathematical object the machine learning system produces, which takes inputs and predicts outputs.
- Feature: a single piece of input information the model uses. "Customer age" is a feature. "Number of website visits last month" is a feature.
- Training: the process of showing the model many examples until it learns to predict well.
- Inference: the process of using a trained model to make predictions on new data.
- Accuracy: the fraction of predictions the model got right. A coarse but useful metric.

- Precision and recall: more careful metrics. Precision is, of the things the model said yes to, what fraction were actually yes. Recall is, of the things that were actually yes, what fraction did the model catch.
- Overfitting: when a model has memorised its training data rather than learning generalisable patterns. The single most common mistake in machine learning.
- Bias: when a model's predictions systematically favour or disadvantage certain groups. Usually traceable to bias in the training data. Often the most important non-technical issue with a deployed model.

How AI is showing up in industries

A short, honest tour. The rules of physics for AI are similar across industries. The implementation realities are not.

Healthcare

AI in medical imaging is now genuinely better than the average radiologist at detecting certain cancers in mammograms and at flagging diabetic retinopathy in eye scans. AI in clinical decision support, drug discovery, and patient triage is more promising than proven. Regulation in healthcare is appropriately heavy. Move carefully. Test extensively. Never deploy a model that affects patient outcomes without a clinician in the loop.

Finance

Credit scoring, fraud detection, algorithmic trading, and customer service have all been transformed. The trade-offs are around fairness and explainability. A loan-denial model that cannot explain why is a regulatory and ethical problem. The fix is to use models that produce interpretable outputs, or to layer an explanation system on top.

Retail and e-commerce

Recommendation systems, demand forecasting, dynamic pricing, supply chain optimisation. The boring middle-of-the-business applications. Often the highest return on investment because the data is plentiful and the metrics are clean.

Manufacturing

Predictive maintenance is the killer application. Sensors on machines, models that predict failure days or weeks ahead, scheduled maintenance instead of emergency repair. Computer vision for quality control is the close second.

Transportation

Autonomous vehicles are still further away than the hype suggests. Routing, scheduling, and fleet optimisation are not. The most exciting AI in transport in 2026 is the boring kind that saves a logistics company a few percent on fuel.

A composite case

A board member at a manufacturing company I advised was responsible for approving the company's AI strategy. He sat through a vendor pitch that used the words "artificial intelligence," "machine learning," "deep learning," "neural network," and "agentic" interchangeably across forty minutes. He approved a two-million-dollar contract.

Six months in, the project was struggling. He asked me to take a look. The system the vendor had delivered was a logistic regression model, the same statistical technique that had existed for fifty years. It was a perfectly competent regression. It was not, by any reasonable definition, deep learning or artificial intelligence or any of the other terms the vendor had used. The board member had paid a deep learning price for a logistic regression.

When I explained this to him, he was angry. Not at the vendor, who he conceded had probably been technically careful in the contract language. At himself, for not knowing the vocabulary well enough to ask the right questions. The vendor had not lied. The vendor had used industry terms in a way that was technically defensible and practically misleading, and he had not had the language to push back.

The company eventually renegotiated the contract. The renegotiation, and the legal time it took, cost them more than the difference between what they paid and what the work was worth. The vocabulary lesson, the board member told me, was the most expensive twenty minutes of education he had ever received.

Objections you will hear

"Leaders do not need to understand this. That is what we hire engineers for."

True if you trust your engineers absolutely and have no vendors to evaluate. False otherwise. Leaders make decisions about which projects to fund, which vendors to hire, and which claims to believe. Sloppy vocabulary in those decisions costs real money. The fifteen minutes it takes to learn the distinctions in this chapter saves more than fifteen minutes per quarter for the rest of your career.

"These distinctions are pedantic. AI is AI."

If AI is AI, then deep learning, logistic regression, and a hand-coded rule engine are equivalent, which they are not. The price difference between them in vendor contracts is sometimes a factor of ten. The capability difference is even larger. The pedantry is the discipline.

"Our team uses these terms loosely. It works fine."

It works fine until a vendor uses them loosely in the other direction. Loose language is fine within a team that shares context. It is dangerous in any conversation that crosses the boundary of the team, including every conversation with a vendor or external consultant.

"The vocabulary changes every six months anyway."

The buzzwords change. The underlying categories do not. AI, machine learning, deep learning, and the difference between supervised and unsupervised have been stable terms for thirty years. The 2026 additions (LLMs, agents, RAG) are stable enough to last the next decade. The investment in learning them pays back over years.

Questions for your leadership team

10. Can each member of our leadership team correctly explain the difference between AI, machine learning, and deep learning to someone who does not know?
11. When we last reviewed a vendor proposal, did we catch any term misuse? Did we even check?
12. Do we share a glossary across the leadership team, or does each of us use these terms slightly differently?
13. Which of our recent AI decisions might have gone better if we had been more precise about vocabulary?
14. If a vendor uses the word "agentic" in our next pitch, who in the room can ask the follow-up questions that will tell us if it means what we think it means?

IF YOU REMEMBER NOTHING ELSE FROM THIS CHAPTER

AI, machine learning, and deep learning are nested boxes, not synonyms. Use them precisely.

Supervised learning is the workhorse. Unsupervised

explores. Reinforcement learns by doing.

Overfitting and bias are the two failure modes that matter most. Ask about them in every project review.

ONE-LINE TAKEAWAY

Learn the vocabulary precisely, even if you never train a model yourself. Sloppy language about AI leads to sloppy decisions about AI.

CHAPTER 3

Where the Cloud Fits

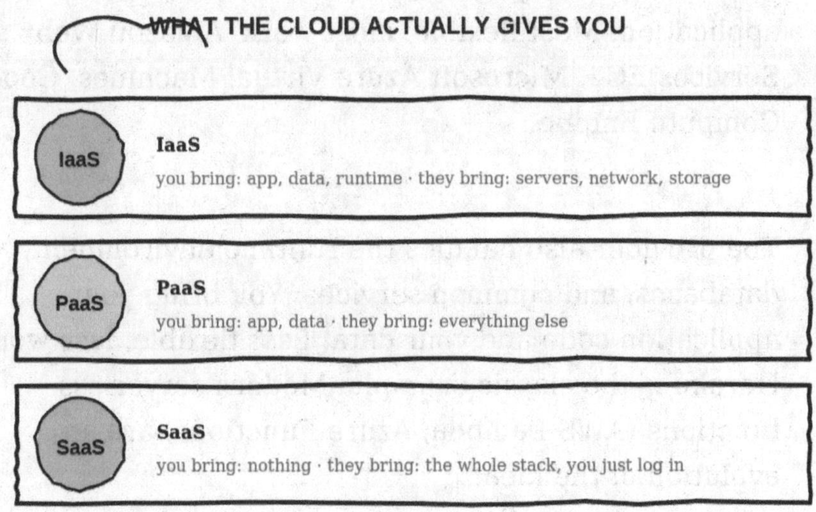

If artificial intelligence is the engine of modern data work, cloud computing is the building it runs in. Almost no serious AI work happens on local servers anymore. Understanding where the cloud fits, how the major providers differ, and which decisions actually matter is part of the modern leader's job.

What the cloud actually gives you

Strip away the marketing and cloud computing is three things: someone else's computers, available on demand, rented by the hour. The three letters everyone uses (IaaS, PaaS, SaaS) describe how much of the stack the provider handles versus how much you do.

Infrastructure as a Service (IaaS)

The provider gives you raw computing capacity: virtual machines, storage, networking. You install everything else, the operating system, the database, your application. Most flexible, most work. Amazon Web Services EC2, Microsoft Azure Virtual Machines, Google Compute Engine.

Platform as a Service (PaaS)

The provider also handles the runtime environment, databases, and common services. You bring your application code and your data. Less flexible, less work. Heroku is the classic example. Modern serverless functions (AWS Lambda, Azure Functions) are an evolution of the idea.

Software as a Service (SaaS)

The provider runs the whole application. You log in and use it. Salesforce, Microsoft 365, Slack, Notion. The customer-facing surface of the cloud, where most non-technical employees actually interact with it.

The choice between these layers is usually misframed as "which is best." It is actually "which layer of the stack do we want to own?" Owning more means more control and more work. Owning less means faster delivery and more vendor dependency. Most organisations end up using all three: SaaS for productivity, PaaS for new development, IaaS for legacy migration and specialised workloads.

Deployment models: public, private, hybrid

Where the cloud physically lives is the second axis.

- Public cloud: the provider's data centres, shared infrastructure, lowest cost, most features.
- Private cloud: dedicated infrastructure either on the provider's hardware or in your own data centre, higher cost, more control.
- Hybrid: a deliberate mix of both, usually with sensitive data kept private and computational scale rented from public when needed.

Most enterprises end up hybrid, not because they planned to but because they could not move everything at once. Plan for hybrid from the start and the transition is easier.

The three big providers, briefly and honestly

There are more than three, but there are three that matter for almost every enterprise decision. A short, opinionated comparison.

Amazon Web Services (AWS)

The most mature. The most services (over two hundred, last I checked). The biggest market share. The steepest learning curve. AWS feels like a hardware store, vast, slightly intimidating, and you can find a tool for anything if you know where to look. Best for organisations with deep cloud expertise or willing to build it. Documentation is excellent. Pricing is opaque.

Microsoft Azure

The most enterprise-friendly. Tight integration with Microsoft's existing enterprise tools (Active Directory, Office, SQL Server). Strongest in regulated industries where the buyer is already a Microsoft shop. AI and ML services are increasingly competitive, particularly since the OpenAI partnership. Best for organisations already running Windows and Office at scale.

Google Cloud Platform (GCP)

The smallest of the three by market share, the strongest by some technical measures (BigQuery, Kubernetes, ML tooling). Best for organisations doing significant data analytics or machine learning work, or those that already use Google Workspace. Smaller ecosystem of third-party tools and integrations than AWS or Azure.

How to actually make the choice

The mistake leaders make is asking "which cloud is best?" That is not the right question. There are three better questions.

15. Which cloud do my people already know? The cost of switching skill bases is enormous and often invisible.

16. Which cloud do my customers and partners use? Integration is dramatically easier inside one ecosystem than across two.
17. Which cloud has my non-negotiable services? If you have a specific workload that runs much better on one provider, that workload may decide for you.

Beyond that, the three big clouds are more alike than different for most workloads. The decision is rarely about which is technically superior. It is about ecosystem fit, talent availability, and existing commitments. The most expensive cloud decision is the one you reverse two years later.

Multi-cloud: the seductive trap

Many organisations talk about being "multi-cloud" to avoid vendor lock-in. The intent is reasonable. The execution is almost always worse than promised.

Genuinely portable multi-cloud requires either restricting yourself to the lowest-common-denominator services (which gives up most of what makes each cloud useful) or building a complex abstraction layer that costs more than the lock-in you were avoiding. For most organisations, a clear primary cloud with a deliberate secondary for specific workloads is the honest pattern. "Multi-cloud" as a strategy slogan often means "we have not chosen yet."

> "Pick a primary cloud. Get good at it. Use a second one only for specific workloads where the math justifies the complexity."

A composite case

A regional bank I worked with had what they called a "multi-cloud strategy." In practice this meant they used AWS for some workloads, Azure for others, Google Cloud for a few experiments, and were paying enterprise licenses on all three. The official reason was vendor diversification. The actual reason, which nobody had wanted to say out loud, was that different VPs had championed different clouds at different times, and no one had ever consolidated the decisions.

Their engineering team supported all three. Their security team had to audit all three. Their procurement team had three vendor relationships. The cloud bill across all three was approximately forty percent higher than a single-cloud deployment of the same workloads would have been. Worse, their engineers were spread thin enough that they were not deeply skilled at any of the three.

A new CIO arrived in 2024 and made the unpopular decision to consolidate. She picked Azure as the primary, on the grounds that the bank's Microsoft licensing was already deep, the regulatory tooling on Azure was strongest for their industry, and the talent market in their city was easiest to hire from.

The consolidation took fifteen months. It was painful. Several engineers left rather than retrain. One vendor relationship had to be wound down carefully because of contractual minimums. The political conversations with the VPs who had championed the other clouds were uncomfortable.

The result, eighteen months after the decision: the cloud bill was thirty-seven percent lower, security incidents were down because the team finally had time to monitor properly, and feature delivery accelerated because the team was now deeply expert in one platform rather than thinly spread across three. The CIO told me later that the project had been politically the hardest of her career and operationally the most rewarding.

Objections you will hear

"Multi-cloud protects us from vendor lock-in"

In theory. In practice, multi-cloud creates a different kind of lock-in, where you cannot move off any of the clouds you depend on because every workload is on one of them and the integration costs are real. True portability requires either using lowest-common-denominator services (which gives up most of what makes each cloud useful) or building expensive abstraction layers. Most companies that talk about multi-cloud for lock-in protection have not done either.

"We should use the best tool for each job, even across clouds"

Sometimes true for genuinely specialised workloads. Almost always false for general computing. The cost of supporting an extra cloud for one workload is usually larger than the benefit of that workload running on the supposedly best platform. The exception is when one cloud has a service that genuinely has no equivalent elsewhere. That exception is rarer than vendors will tell you.

"Our regulators require multi-cloud"

Sometimes true for specific regulated industries, especially financial services in some jurisdictions. Usually overstated by people who do not want to make a decision. Check the regulator's actual requirement. It is often "redundancy across geographic regions," which a single cloud can provide, not "different cloud providers."

"Our CEO has a relationship with one vendor we cannot ignore"

Vendor relationships are real and political and cannot be wished away. They also should not decide your architecture. If the CEO's relationship is genuinely the reason you are on a cloud that is otherwise the wrong fit, that is a conversation to have with the CEO, not a constraint to silently accept. The cost of an architecture chosen for political rather than technical reasons compounds for years.

Questions for your leadership team

18. Do we have a documented primary cloud, or are we drifting across multiple by default?
19. What is the actual annual cost, including engineering overhead, of our cloud complexity?
20. If we standardised on one primary cloud, which workloads would have to move? What would the migration cost be, and how does that compare to the ongoing cost of not consolidating?
21. Are our engineers deeply skilled at one cloud or thinly spread across several?

22. When was the last time we re-evaluated our cloud strategy against current needs, rather than defending past decisions?

IF YOU REMEMBER NOTHING ELSE FROM THIS CHAPTER

The cloud is rented computing. Three layers (IaaS, PaaS, SaaS) describe how much you rent.

Three providers dominate. They are more alike than different for most workloads.

Pick a primary cloud based on existing skills, partners, and non-negotiable services. Not on which is technically best.

"Multi-cloud" as a strategy is usually code for "we have not chosen yet."

ONE-LINE TAKEAWAY

Choose a primary cloud deliberately and get good at it. The cost of indecision is higher than the cost of vendor lock-in.

PART II

The Generative Era

What LLMs and agents are, what they can and cannot do, and the stack you build on them.

CHAPTER 4

What an LLM Actually Is

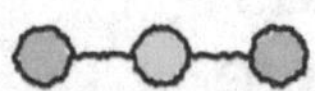

A CEO called me in the autumn of 2023.

He had just paid a consulting firm four hundred thousand dollars for an AI strategy deck. He wanted me to look at it. The deck was eighty slides, beautifully designed, full of phrases like "AI-first organisation" and "cognitive enterprise transformation." I read it twice in his office while he watched me.

Then I asked him one question. "What does an LLM do, in one sentence, no jargon?"

He could not answer.

Neither could anyone else in the room. The consulting firm that had written the deck had moved on to the next client. The four hundred thousand had bought a glossary, not an understanding.

This chapter exists because of that meeting. A leader who cannot describe an LLM in one plain sentence is going to make bad decisions about LLMs no matter how thick the strategy deck. The fix is not another deck. The fix is fifteen minutes of clear explanation, which is what this chapter tries to give you.

In one sentence, no jargon

Here is the sentence I gave that CEO, and the one I will give you.

> "A large language model is a very large pattern-matcher that has read most of what humans have ever written, and uses that reading to predict what word should come next."

That is the whole trick. It is not thinking. It is not understanding in any deep sense of the word. It is a piece of mathematics, trained on roughly all the text humans have produced, that has gotten extraordinarily good at predicting what word should come next given the words that came before.

Everything an LLM does, every essay it writes, every line of code it generates, every question it answers, is the same operation underneath. Take the words so far, predict the next word, append it, repeat. The miracle is that this simple operation, run at sufficient scale on sufficient data, produces output that feels like conversation. The illusion is so good that even the people who build these systems sometimes forget what is happening under the hood.

Hold that picture in your head. A very large statistical engine for guessing the next word. Everything that follows in this chapter is a footnote to that one sentence.

Why the trick works

The first time I saw GPT-2 generate a paragraph in 2019, I thought it was a parlour trick. The grammar was right. The sentences were mostly coherent. But the content was nonsense. It would tell you confidently that George Washington had invented the bicycle. It was a sophisticated mimic, not a useful tool.

Then GPT-3 happened in 2020. Then GPT-4 in 2023. Then a parade of models from a parade of companies. Somewhere along the way the mimicry got so good that it stopped being mimicry. The model that confidently invents nonsense is now the model that drafts your contracts, debugs your code, and writes the first version of your kid's biology essay. And the underlying trick did not change. We just scaled it.

What scaled, specifically, was three things. The amount of text the model trained on, from millions of pages to trillions. The size of the model itself, from millions of parameters to hundreds of billions. And the amount of compute used to train it, by factors of thousands. Each of these jumped together, each pushed the others, and at some threshold around 2022 the output crossed from "interesting research" to "useful product."

No one knows exactly why crossing that threshold produced the qualitative jump we got. There are theories. There is honest mystery. The honest version is that nobody set out to build a thing that could write a passable college essay, and the people who built it are themselves not quite sure how it turned out to do that. Hold that in mind too. The systems you are now deciding to deploy at your company were not designed in the careful way a database is designed. They were grown, in a sense, by feeding them a planet's worth of text and watching what emerged.

The model landscape, briefly

Naming names. By the time you read this, the lineup will have shifted. New entrants. New version numbers. Old names quietly retired. But the categories will hold, so the names below are useful as anchors, not as a buying guide.

At the top of the market in 2026 are the frontier closed models. OpenAI's GPT line. Anthropic's Claude. Google's Gemini. These are the most capable models. They are also the most expensive, both per query and as a strategic dependency. You do not own them. The vendor can change the model, the price, the terms, the safety filters, at any time, and your product changes with them.

Below the frontier are the open weights models. Meta's Llama. Mistral's various releases. A growing number of open-source projects. These are not quite as capable as the top closed models, but the gap has narrowed dramatically. The trade-off goes the other way. You can run them on your own infrastructure. You control the weights. You can fine-tune them on your private data without sending that data to a vendor. The cost is operational complexity.

Below those are the specialised small models. Models trained or fine-tuned to be very good at one narrow thing. A model that does nothing but extract structured data from medical notes. A model that does nothing but translate between two specific programming languages. These are often the right choice when the use case is bounded, because they are cheaper, faster, and more controllable than the frontier.

If you are making a decision about which model to use, the questions to ask, in order:

- How sensitive is the data the model will see? If it is highly sensitive, you may need to keep the model on your own infrastructure, which pushes you toward open weights.
- How varied is the work? If the model needs to do many different things, you want a frontier model. If it needs to do one thing very well and very cheaply, a specialised small model often wins.
- How important is consistency over time? Closed frontier models can change underneath you. Open weights models do not, unless you choose to change them.
- How much can you spend per query? Frontier models in 2026 are roughly ten to a hundred times more expensive per query than small models. At low volumes this is irrelevant. At high volumes it determines whether the use case has positive unit economics.

The three ways to put an LLM to work

This is the part of the chapter most leaders need most, because the choice between these three modes is the choice that decides whether your AI project succeeds or wastes a year of your engineering team's time.

Prompting

The simplest mode. You give the model an instruction in plain language. It responds. "Summarise this report in five bullet points." "Translate this support email into Japanese." "Draft a friendly response to this customer complaint." The model does the work. You read the output. You ship it or you edit it.

Prompting is enough for a startling number of use cases. Probably most of the ones you are considering. The mistake to avoid is reaching for the more complex modes when prompting alone would have been fine. I have watched companies spend nine months building elaborate fine-tuning pipelines for problems a careful prompt would have solved in an afternoon.

The skill in prompting is mostly about specificity. The model is better than you think at following instructions if you give it good ones. "Write a marketing email" is a bad prompt. "Write a 120-word marketing email to existing customers of a coffee shop, announcing a new seasonal blend, in a warm and slightly self-deprecating tone, with a clear call to action to visit the shop this Saturday" is a good prompt. The difference in output is dramatic.

Retrieval-augmented generation, or RAG

The mode for when the model needs to know things it was not trained on. Your company's internal policies. Last month's customer support tickets. The text of every contract you have signed. The model itself does not know any of this. You give it that knowledge at query time.

The mechanism is straightforward in concept and a little fiddly in practice. You take your private documents, break them into chunks, convert each chunk into a numerical fingerprint, and store the fingerprints in a database that can find them quickly. When a user asks the model a question, you first find the chunks of your private data that look most relevant to the question, then you hand those chunks to the model along with the question, and tell the model to answer the question using the chunks.

The model now has access to your private knowledge without having been retrained on it. It will cite the source documents. It will say "I do not know" when the documents do not contain the answer, if you tell it to. It will not leak your private data into its general training, because no retraining happened.

RAG is the dominant pattern in 2026 for enterprise AI work. Most of the LLM applications you will build in the next two years will look like some flavour of this. Get the pattern right and your team will move quickly. Get it wrong, with bad chunking or bad retrieval or bad prompts to the model, and you will spend a year wondering why the AI keeps making things up.

Fine-tuning

The third mode, and the one most often used when one of the first two would have been enough.

Fine-tuning means taking a pre-trained model and training it further on your specific data. The model adjusts its internal weights to be better at your particular task. After fine-tuning, the model knows your domain in a way it did not before. It uses your terminology. It follows your style. It handles edge cases specific to your business.

This sounds appealing. In practice, most teams should not reach for it until they have tried prompting and RAG first. Fine-tuning is expensive to do, expensive to maintain, and creates a long-term dependency on a model version you now have to keep up to date as the base model improves. There are good reasons to fine-tune. Strong domain-specific terminology that prompting cannot consistently teach. A specific output format the model keeps getting wrong. A latency requirement that means the prompts have to be short, which means the instructions have to be baked into the model itself.

But most of the time, when a leader asks "should we fine-tune?", the honest answer is "probably not yet." Try prompting. Try RAG. Measure where each falls short. Then, if you still need it, fine-tune the specific gap. Do not start with fine-tuning. It is the most fashionable answer and rarely the right first move.

Cost, and why everyone is wrong about it

In 2023, calling a frontier LLM cost several cents per query. By 2026, the cost has fallen by roughly a factor of ten, and continues to fall. By 2028 it will fall again by another factor of ten if current trends hold.

This has two implications most leaders have not fully internalised.

First, the use cases that are uneconomic today will be economic next year. If your team analyses a use case in

January and concludes the cost per query makes it unworkable, do not file the analysis away. File a reminder to redo the analysis in nine months. The cost curve is steep enough that what was impossible last year is often trivially affordable now.

Second, the use cases that look cheap can suddenly look expensive when they succeed. A customer support assistant that costs a tenth of a cent per query is cheap until customers love it and use it a hundred times per month each, at which point it is real money. Always run the cost model not against today's volume but against tomorrow's. The companies that get burned here are not the ones that fail to launch. They are the ones that succeed too well, too fast.

When not to use an LLM

I want to spend a section on this because it is the most under-discussed topic in the entire field, and the one I most often have to talk leaders down from.

LLMs are bad at math. Not arithmetic, math. They can do simple sums correctly most of the time, but for anything involving careful calculation, you want a calculator or a spreadsheet or a piece of actual code, not a model that is guessing what number comes next.

LLMs are bad at retrieval of exact facts. They will confidently tell you the wrong date, the wrong number, the wrong quote. For anything where the answer is a specific known fact, you want a database lookup, not a model that is hallucinating something plausible.

LLMs are bad at multi-step reasoning that requires keeping precise state. Plan a complex multi-step process, with branching decisions, and the model will lose the thread halfway through. Use a real workflow engine for that, with the model handling the natural-language parts inside the steps.

LLMs are bad at being current. The model's knowledge has a cutoff date, often months or years before you are using it. For anything where currency matters, you have to feed the model the current information at query time, or use a different tool entirely.

And LLMs are bad at being legally accountable. If a person says something defamatory, you can sue the person. If a model says something defamatory, the legal answer is currently "unclear, expensive, and slow." For anything where accountability matters, you want a human in the loop.

In all these cases the right answer is not "do not use AI." The right answer is "do not use an LLM for this part of the problem." Pick the right tool for the job. The leaders who reach for an LLM for every problem they encounter, because LLMs are the fashionable tool of the moment, are the same leaders who tried to use blockchain for everything in 2018 and the cloud for everything in 2012. They tend to spend a lot of money and ship things that do not work.

A composite case

A healthcare services company I worked with in 2024 wanted to use AI to summarise patient call transcripts for the clinicians who saw the patient next. The use case was real. Clinicians wasted ten or fifteen minutes per patient reading through call notes before each visit. Cutting that to two minutes would save thousands of clinician hours a year.

The first version they built tried to use a frontier model with a very long prompt that included the company's clinical guidelines, the patient's history, and the new transcript. It worked, but it was slow and expensive, about a dollar per summary at the volume they expected. The math did not work.

The second version was RAG. They kept the company's clinical guidelines in a vector store, retrieved only the relevant guidelines for each patient, and gave the model a much shorter prompt. The cost dropped by a factor of eight. The summaries were comparably good. They were preparing to ship.

Then the clinical safety team came in and pointed out something nobody had thought of. The model had no way to know which of its summaries were trustworthy and which were guesses. A clinician reading a summary that was confidently wrong could make a bad treatment decision. The cost of being wrong, even occasionally, was a malpractice lawsuit.

The third version was the one that shipped. It used RAG for the routine summaries, but it also flagged any sentence in the summary where the model's confidence was low, and any claim that did not have direct support in the transcript. Those sentences were shown to the clinician in yellow, with a note to verify. The clinician's job became reading a short summary with the uncertain parts highlighted, rather than reading the whole transcript or trusting an opaque AI summary blindly.

The third version was slower and more expensive to build than the first two. It was also the only one that was actually safe to deploy in healthcare. The lesson the team took away, and that I keep coming back to in every leadership conversation I have about LLMs, is this. The hardest part of the LLM project is almost never the LLM. It is the part of the system around the LLM that decides what to do when the LLM is wrong.

Objections you will hear

"LLMs hallucinate. They are unreliable."

True. They will confidently say wrong things. The fix is not to expect them to stop. The fix is to design the system around them so the hallucinations are caught before they do harm. Retrieve sources. Show confidence levels. Put humans in the loop where the stakes are high. The reliable systems being shipped in 2026 are not reliable because the LLMs got perfect. They are reliable because the engineers around the LLMs got smart about handling imperfection.

"This is just a fad. The hype cycle will pass."

The hype cycle will pass. The technology will not. We are at the stage of LLMs that the public internet was at in 1998. There was a lot of hype, much of which was silly. There was also a real technology underneath that proceeded to change everything. The difference between the two is hard to see at the time and obvious in retrospect. I would not bet against the underlying trend.

"We cannot let our data go to an external model"

In 2023 this was a stopper. In 2026 it is rarely one. You can run open-weight models on your own infrastructure. You can use cloud providers' enterprise tiers with strict data isolation. You can negotiate contracts that prohibit training on your data. The path exists. The companies that say they cannot use LLMs for data reasons are usually really saying they have not done the work to figure out how.

"Our use case is too specialised. The general model will not work."

This is often partially true but rarely a stopper. Specialised use cases benefit from RAG or fine-tuning. They almost never require building a model from scratch, which would be enormously expensive and almost always worse than starting from a frontier model. If a vendor is trying to sell you a custom model for a use case that could be solved with prompting and RAG, get a second opinion.

Questions for your leadership team

23. Can every person on our leadership team describe an LLM in one plain sentence?
24. For each LLM use case we are pursuing, have we explicitly chosen prompting, RAG, or fine-tuning? Why that choice?

25. Have we identified the failure modes (hallucination, currency, math, accountability) that matter for our use case, and designed the system around them?
26. Have we modelled the cost not at today's volume but at tomorrow's success volume?
27. For each use case, is the LLM the right tool, or is something simpler (a database, a calculator, a workflow engine) the better answer?
28. Do we know which LLM provider we are betting on, what would happen if they changed terms tomorrow, and what our exit plan is?

SIX THINGS TO TAKE FROM THIS CHAPTER

An LLM is a very large pattern-matcher that predicts the next word. Hold that picture in your head.

Three modes to choose from: prompting, RAG, fine-tuning. Try them in that order. Most teams skip ahead and regret it.

Frontier closed models are most capable. Open weights models are more controllable. Specialised small models are cheaper. Pick deliberately.

LLM cost per query is dropping roughly tenfold per year. Re-evaluate "uneconomic" use cases every nine months.

Know when not to use one. Math, exact facts, multi-step reasoning, currency, accountability. Pick the right tool for the job.

The hardest part of an LLM project is almost never the LLM. It is designing what happens when the LLM is wrong.

ONE-LINE TAKEAWAY

A leader who cannot describe an LLM in one plain sentence is going to make bad decisions about LLMs. Spend the fifteen minutes to understand the pattern-matcher underneath. Choose the right mode for the use case. Design the system to handle the model being wrong. Re-check your cost assumptions every quarter. Everything else is detail.

CHAPTER 5

What the Machine Cannot Do

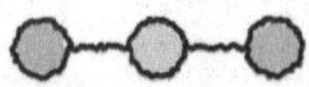

Two summers ago I sat with an engineer I have known for fifteen years.

We were debugging a customer issue, the same kind of work I had watched her do a thousand times. Except this time, before her fingers touched the keyboard, she opened a small chat window in the corner of her editor and described the bug in plain English. The window thought for a moment. Then it wrote a fix. She read it, frowned, asked it to try again. The second version was right. The whole thing took six minutes.

I had not seen her debug a problem in under an hour in fifteen years.

We did not speak about it that afternoon. We just kept working. But on the train home I sat looking at my own reflection in the window and realised that something I had assumed about my profession was no longer true. The thing I had spent twenty years getting good at, the thing I had taught dozens of younger engineers to do, the thing my profession was organised around, the slow patient act of typing code by hand, was being quietly replaced by something else. And what bothered me most was not the loss. It was that I could not yet say what was being kept.

This chapter is the answer I have been working towards since that train ride. What changes when machines can write code. And, more importantly, what does not.

Start at the end

I want to put the punchline first, because every paragraph that follows will only make sense if you read it through this lens.

Machines can now write code. They cannot decide what is worth writing. They cannot tell you whether the thing your customer asked for is actually the thing they need. They cannot stand in front of your board and explain why a system failed at three in the morning. They cannot fall in love with a problem and keep coming back to it for a decade until they finally crack it. They cannot do any of the work that a real engineer does after the typing is done.

> "Vibe coding can build your code in a day. Creativity and intuition are given only to a human."
>
> Sarah

If you remember nothing else from this book, remember that sentence. Every leader I have watched make a serious mistake about AI in the past two years made it because they confused the typing with the thinking. They fired engineers, or refused to hire them, because they believed the machine had made the work obsolete. A year later most of those leaders were quietly trying to hire those same engineers back, often at premium prices, because the work had not gone obsolete. Only the typing had.

Hold that thought. Now we can talk about what actually changed.

What actually changed

For about forty years, the way you wrote software was the way you wrote software. You sat at a keyboard. You typed characters. You read documentation when you got stuck. Every few years some new editor or language or framework made the typing slightly more pleasant, but the act itself was unchanged from the days of punch cards. My grandmother could have understood my workday.

Then, very quickly, the act stopped being primary.

It began with GitHub Copilot in 2021, which finished your sentences. By 2024 there was Cursor, which would write whole functions if you asked it to. By 2025 there was Claude Code, which would happily write whole features across multiple files. Windsurf, Cline, Aider, half a dozen others. The names will be different by the time you read this. The category will not.

(I am going to name specific tools throughout this chapter because that is more useful to you than the consultant-speak of "AI coding assistants." Treat the names as illustrative. The category outlasts every particular product in it.)

Here is what is true at every company I have advised that has made the transition seriously.

A good senior engineer is now doing about twice as much work in a day as they were eighteen months ago. Sometimes three times. Not because they are working harder, they are working less. The boring work, the tests, the refactors, the documentation, the glue code between two systems, the new endpoint that looks like the last forty endpoints, all of that has been compressed from days into hours. From hours into minutes.

The hard work, on the other hand, takes about as long as it ever did. Designing a system that will live for ten years. Writing the code that handles the credit card transaction or the medical record or the autonomous vehicle's emergency brake. Optimising the path through a database that runs a million times a second. Choosing which feature is worth building at all. None of this got faster. Some of it, honestly, got slower, because the engineer now has to also review the AI's suggestions and decide which to trust.

So the curve of an engineer's day has changed shape. The boring middle has collapsed. The hard parts at either end have not. This is the entire story, and most of what follows is footnotes.

The strange phrase that stuck

Sometime in 2024 the practice acquired a name, which is how you know it has become real. People started calling it vibe coding.

I disliked the phrase when I first heard it. It sounded glib. The work it described was not glib. It was a precise, demanding discipline that took most engineers a year to get good at. But the phrase stuck, and the more I watched engineers do it, the more I understood why.

Vibe coding is what happens when you stop telling the machine exactly what to do, character by character, and start telling it what you want, sentence by sentence. You describe the thing. The machine attempts the thing. You read what it wrote. You correct it. It tries again. The cycle is faster than typing. It is also, if you are honest about it, more interesting. The engineer's job becomes specifying, judging, and refusing, more than producing.

Watch a competent vibe coder for an hour and you will see something that looks more like a writer editing a draft than a typist transcribing a recording. The intelligence is in the choices. The choices are still entirely human.

That last sentence is the one I keep coming back to. The intelligence is in the choices. The machine can produce a great deal of plausible code. The choice of which plausible code is actually correct, which is actually good, which is actually worth shipping, lives in a human head and probably always will.

Where the machine is excellent

I have given up trying to be exhaustive about this. The list gets longer every quarter. But the categories where the machine is now genuinely excellent are:

- Writing the first ugly draft of any piece of code. The blank page is dead. There is no longer such a thing as not knowing where to start.
- Translating between systems. Take this JSON, give me the SQL. Take this code, give me the equivalent in another language. Take this messy log file, parse it for me. These tasks used to consume hours. They now take seconds.
- Writing tests. Engineers used to skip tests because tests were boring. Now the AI writes them happily. Test coverage in the codebases I see has roughly doubled in two years, and the codebases are demonstrably more reliable for it.
- Writing documentation. Same logic. The thing engineers always meant to do, the thing they never quite got around to, the AI will do without complaining.
- Reading unfamiliar code. "Explain what this function does and why" is a query the AI is genuinely good at. New engineers ramp up on a strange codebase in a fraction of the time it used to take.

None of this is glamorous. All of it is the work that ate the better part of an engineer's week before. Getting all of it back is enormous.

Where the machine is still bad, and probably always will be

And here is where the chapter's opening sentence comes home to roost.

The machine cannot tell you what to build. It cannot stand in a room with three product managers, a designer, and a CEO, listen for an hour, and know in its body that one of the proposed features is the right one and the other two will sink the quarter. That knowledge is not in any training corpus because it is partly emotional, partly political, partly aesthetic, and partly born of having been wrong about exactly this kind of decision many times before.

The machine cannot make a design taste right. It can give you ten variations of an interface. It cannot tell you which of them will make a tired person at the end of a long day smile. That smile is the entire business in many companies, and it is decided by humans with eyes and hearts.

The machine cannot notice when something feels off. The best engineers I have ever worked with have an instinct, before they can articulate it, that a piece of code is wrong. They cannot tell you why. They just know. Then they read it slowly and find the bug. That instinct is the residue of a thousand bugs caught the hard way. The machine has none.

The machine cannot take responsibility. If a system you built fails at three in the morning, a human gets paged. A human makes the call about whether to roll back, whether to wake the CEO, whether to tell the customer.

A human stands in the review afterwards and says, yes, I did that, here is what I learned. No model can hold that accountability. No model should.

And the machine cannot build a relationship. The hardest part of senior engineering, the part nobody told me about when I was twenty-five, is almost never the code. It is convincing a stakeholder to change scope. It is mediating a fight between two teams that both think they own the same system. It is patiently teaching a junior engineer the thing you wish someone had patiently taught you. AI does none of this. It will never do any of this. The companies that figure this out will treat their engineers as something close to sacred.

The junior engineer question

This is the part of the chapter I have rewritten the most times, because I do not think I have the answer right yet, and the stakes are very high.

In the old world, you became a senior engineer the only way anyone ever became one. You wrote a lot of bad code. You had it reviewed by people more experienced than you. You fixed your mistakes. You wrote some slightly better bad code. You had it reviewed again. You did this for about seven years, and at the end you had instincts. Not knowledge, instincts. The instincts were what made you senior.

In the new world, juniors can ship working code on their first day. Whatever they cannot do themselves, the machine fills in. The output looks good. The customer is happy. The product manager is happy. The juniors are happy. The instincts, however, are not being formed, because nothing has been struggled with.

I do not know how this ends. I am genuinely unsure. The honest companies I talk to are unsure too.

Some have started requiring juniors to write the first version of any piece of work without AI assistance. The struggle is preserved deliberately. The AI becomes a kind of teacher rather than a substitute. I think this is probably right, though I worry about what happens when these juniors leave for jobs where the discipline is not enforced.

Some have stopped hiring juniors altogether. The math seems to work in the short term, a senior engineer plus AI is more productive than a senior plus a junior. The math does not work in the long term. If everyone does this, in ten years there will be no seniors, because there were no juniors to grow into them. I think this is the gravest unforced error in the industry right now, and I would not bet on the companies making it.

Some have invented entirely new junior tracks. Less focus on coding speed, more on system understanding, code reading, architectural pattern recognition. I am cautiously hopeful about this approach, though it is too early to know if it produces the same depth in the long run.

If I had to advise a CTO today, I would say this. Keep hiring juniors. Pay them slightly more than you used to, because there will be fewer of them and the ones who can think will be worth it. Give them work that is genuinely hard, and require them to engage with the difficulty before reaching for the machine. And take their long-term development seriously enough that you measure it in years, not quarters. You are building the seniors of 2035. Nothing the machine does changes that responsibility.

Five decisions a leader actually has to make

I have tried to keep the structural scaffolding to a minimum in this chapter, but here we need a list, because these are concrete decisions that I see leaders fail to make, and the failure costs them money or talent or both.

Which tools, and who pays

Pick one or two tools across the organisation. The differences between Cursor and Claude Code and Copilot and Windsurf matter less than the consistency of supporting one of them properly. Pay for it centrally, do not make engineers expense their own. Per-seat pricing in 2026 is roughly ten to forty dollars a month. For a hundred engineers, you are looking at twelve to forty-eight thousand dollars a year. This is rounding error against the productivity gain. Just budget for it.

What is and is not allowed to touch sensitive code

Write a policy. The simple version: AI assistance is allowed everywhere by default. AI assistance is restricted on code that handles credentials, payments, authentication, regulated user data, and anything the legal or security team has flagged as crown-jewel intellectual property. Engineers can request exceptions and they go through a defined path. This sounds bureaucratic. It is the difference between sleeping at night and not.

How code review is done now

The single most dangerous pattern in 2026 engineering is the rubber-stamp review of AI-generated code. The diff is large, the author trusts the AI, the reviewer trusts the author, no one really reads it. This is exactly how subtle bugs and security holes shipped at three companies I have advised this year alone. The cultural mitigation, said out loud and often, is that AI-generated code gets more careful review, not less. The technical mitigation is to surface in the review tool whether a diff was AI-generated, so the reviewer knows what they are looking at.

The intellectual property question

The Copilot lawsuit is still being worked out as I write this. The honest answer is that there is some unresolved legal risk that AI-generated code may turn out to be partly derived from open-source training data with viral licenses. Most vendors now have indemnification clauses for this. Have your legal team read them. Have your legal team also read the data-handling clauses, because what happens to your code when it goes through the AI's servers matters.

The hiring plan

Your hiring plan from 2021 is probably wrong. The shape of the team has shifted. You need fewer generalist mid-level engineers, more strong seniors, more specialists in security and architecture and performance, and a genuinely thought-through approach to juniors as discussed above. If your job postings still look like they did three years ago, your money is being spent on the wrong shape of organisation.

A composite case

A mid-sized SaaS company I worked with rolled out Cursor across eighty engineers in three months in 2025. Their CTO had spent twenty years in the industry and was, in her own description, deeply suspicious of the technology. That suspicion turned out to be useful.

She insisted on a pilot. Fifteen senior engineers, six weeks, before anyone else got access. The pilot had three jobs: measure what actually changed, find the failure modes nobody was talking about, and write the policies the broader rollout would need.

Six weeks later, the data was both exciting and sobering. Productivity on routine work was up about thirty-five percent on average. Some engineers, the enthusiastic adopters, doubled their output. Some, the skeptics, saw maybe ten percent gain. On the hard architectural work, the gain was approximately zero. The AI helped with the typing and not with the thinking, exactly as we have been saying.

Test coverage on new code went up from sixty percent to eighty-five percent. The AI cheerfully wrote tests the engineers had been postponing for months. The code was better because of it.

There were also two incidents. An AI-generated database migration script ran in the wrong environment. An AI-generated authentication change had a subtle privilege escalation bug, caught in code review the day before it would have shipped. Both were near misses. Both became case studies in the broader rollout.

Based on the pilot, the rollout included: a written policy on what AI could and could not touch; an updated code review checklist; a weekly internal meeting where engineers shared useful prompts and failure modes; mandatory training on the IP and security implications; and a clear path for engineers to request exceptions on the restricted parts of the codebase.

A year in, the company estimates AI assistance has saved between twenty and thirty percent of engineering hours company-wide. They have redeployed those hours into more features rather than fewer engineers. Two engineers were promoted into new AI engineering roles focused on the internal tooling. The hiring pattern has shifted toward fewer juniors and more strong mid-level hires.

The CTO told me the most important decision they made was the one announcement she made to the whole engineering team on the day Cursor went company-wide. She said, in front of everyone, that no engineer would be laid off because of AI productivity gains. The gain would go into more ambition, not fewer people. She said this for moral reasons, but the operational effect was significant. Engineers stopped being defensive about the tool. They started showing each other the clever things they had figured out. The rollout went twice as fast as the comparable rollouts I had seen elsewhere. The thing that made the difference was a sentence.

Objections you will hear

And how I would answer them, having heard each of these dozens of times.

"AI will eventually replace engineers entirely"

It will not. Every wave of automation in software, from compilers to libraries to cloud computing, was supposed to reduce demand for engineers. Every one of them increased it instead, because the lower cost of writing software expanded what was worth building. AI is the biggest wave but the same dynamic holds. The engineers who use AI well will be in higher demand, not lower. The work that is left for humans is the work that mattered most all along.

"Juniors will never learn the fundamentals"

This is the right concern. The wrong response is to ban AI for juniors. The right response is to preserve the struggle. Give them real problems. Require them to engage with the problem before reaching for the machine. Pair them with seniors who can model the thinking. The juniors who come out of this kind of training in five years will be more interesting engineers than my generation was, because they will have learned to think with a machine and without one.

"AI writes bad code"

Sometimes. Humans also write bad code, often confidently. The mitigation is the same in both cases.

Rigorous review. Good tests. Observability in production. AI-generated code has a different distribution of failure modes than human-generated code, and your team needs to learn that distribution. It is not categorically worse. It is differently flawed.

"We cannot use it for security reasons"

This was a defensible position in 2023. It is rarely defensible now. Self-hosted options exist. On-premise deployments exist. Contract clauses prohibit training on your data. The opportunity cost of not using AI is now a competitive disadvantage that your security argument has to weigh against, not trump.

"It is a fad"

It is not. The technology is meaningfully better than a year ago, and a year ago was meaningfully better than the year before. The trajectory is steep. Betting against it in 2026 is roughly equivalent to betting against the public internet in 1998. Some companies did. Most of them do not exist anymore.

Questions to bring to your next leadership meeting

If you read this chapter and remember none of the specifics but show up at your next meeting and ask these seven questions, you will have done more than ninety percent of leaders are currently doing.

29. Have we said publicly, to our engineers, that AI is for amplifying their work rather than replacing them?

30. Which teams are using AI tools, with what tools, paid by whom, with what approval?
31. What is our written policy on AI assistance for sensitive code?
32. Has our code review process been updated for AI-generated diffs?
33. What is our IP exposure if the legal landscape on AI-generated code shifts against us?
34. How are we developing our junior engineers in a world where they can ship working code on day one?
35. Does our 2026 and 2027 hiring plan reflect the new shape of engineering work, or is it the 2021 plan with a fresh date stamp?

SIX THINGS TO TAKE FROM THIS CHAPTER

Machines write the code. Humans decide what is worth writing. That distinction does not blur.

Vibe coding can finish a feature in a day. Creativity and intuition belong to the human, and always will.

The boring middle of engineering work has collapsed. The hard ends, design, judgement, taste, accountability, have not.

The biggest cultural risk is rubber-stamp code review. Read the diffs. Read them carefully.

The biggest long-term risk is the slow collapse of the junior pipeline. Hire juniors. Preserve the struggle.

Say out loud, in front of your team, that AI is here to amplify them, not replace them. That single sentence will earn you more goodwill than any tool rollout.

ONE-LINE TAKEAWAY

The machine can build the code in a day. It cannot decide what is worth building, notice when something feels wrong, take responsibility when it fails, or fall in love with a problem for ten years. Those are human jobs. They always were. They will remain so for as long as software matters.

CHAPTER 6

Agents Are Not a Miracle

Let me tell you about the brightest fresh graduate I ever hired.

She was twenty-three. Came out of Stanford with a perfect GPA. Had three internships at companies you have heard of. In her first week she solved a problem one of my senior engineers had been stuck on for two months. I was thrilled. I told her so. I started giving her more work, then more, then more, because she made everything she touched look effortless.

Six weeks in, I was so impressed I left for a two-week holiday in Italy. I told her to keep going on the customer onboarding project she was leading. I told the team she was the lead. I sent the customer she was working with a friendly note saying they were in good hands.

When I came back, we had lost two customers.

She had been making decisions she did not have the experience to make, in a domain she did not yet understand, with confidence I had mistaken for competence. The work looked good. The work was good, in the parts of the work she actually understood. But the parts of the work she did not understand, the parts that required ten years of pattern-matching against painful experience, she did not even know she was getting wrong. And neither did I, because I was in Italy.

That is what an agent feels like in 2026. That is the chapter.

What we mean by an agent

An LLM, by itself, is a thing you talk to. You give it words. It gives you words back. The conversation lives inside one window. When the window closes the conversation is over. The LLM cannot do anything in the world without you copying its output and acting on it yourself.

An agent is the same LLM, but with hands.

Specifically, an agent is an LLM placed inside a small loop, given access to a set of tools, and told to keep going until a task is finished. The tools might be the ability to browse a website. To send an email. To run a piece of code. To look something up in your database. To call another API. The loop is the agent looking at the situation, deciding which tool to use, using it, looking at the result, deciding what to do next.

That is the whole picture. An LLM, plus tools, plus a loop. Run it for long enough and the agent can do things that, watching from the outside, feel impossible. It can research a topic across the web for an hour and come back with a summary. It can refactor a codebase across thirty files. It can answer a customer support ticket that required pulling information from four different systems. The combined effect of "LLM plus hands plus persistence" is qualitatively different from the LLM by itself.

This is why agents are exciting. It is also why they are dangerous, and why every leadership team needs to understand exactly where the danger lives.

The Stanford-graduate problem, in full

Hold the analogy in your head while we work through this, because it is the most useful frame I have found for how to think about agents.

My fresh graduate was impressive in narrow ways. She had read more recent papers than my senior engineers. She had used more modern tools. She wrote cleaner code on the first try. She moved faster on well-defined problems than anyone else on the team. All of this was real and worth paying for.

What she lacked was every form of judgement that comes from having been wrong in the past. She did not know which customers were sensitive about pricing changes because she had never had a customer scream at her after a poorly-timed price increase. She did not know which engineers on the team would push back hard if asked to change scope on a Friday because she had not yet been in a Friday-evening fight about scope. She did not know that the procurement team at the customer's company moved slowly in December because she had never lost a quarter to a holiday slowdown. The work she could see, she could do. The work she could not see, she did not know she was missing.

Agents are exactly the same.

> “An agent is a fresh Stanford graduate who happens to be very fast. It can do remarkable work in the parts of the job it understands. It will confidently make mistakes in the parts of the job it does not know exist. And it will not know to ask.”
>
> Sarah

The mistake leaders are about to make with agents is the same mistake I made with my graduate. They will see the impressive work in the well-defined corner of the problem and assume the impressive work generalises to the whole job. They will redeploy the senior people, or fire them, on the strength of what the agent can do in the narrow corner. Then they will go on the operational equivalent of a vacation in Italy, which in modern corporate life means "we have automated this domain, the team moves on to other priorities, no one is watching closely." And then they will discover, six weeks later or six months later, that the part of the job nobody was watching is the part that mattered most.

What agents actually do well in 2026

This is not a chapter about dismissing agents. The technology is genuine and the use cases are real. Here is what I have watched work.

Narrow, well-defined workflows

Customer support that follows a known playbook for a known set of common tickets. The agent reads the ticket, checks the relevant system, drafts the response, hands it to a human for approval. If the ticket falls outside the known playbook, the agent escalates to a human rather than guessing. This works because the boundary of the agent's competence is defined explicitly, and there is a human in the loop for anything outside it.

Research and synthesis where the answer can be checked

An agent that gathers information from many sources and produces a summary is genuinely useful, provided the consumer of the summary can verify it. A sales agent that researches a prospect company before a meeting is useful because the salesperson can sanity-check the output in five minutes. A research agent that produces a literature review is useful because a researcher can verify the citations exist. The pattern that works is fast first draft from the agent, checked by a competent human, before any action is taken.

Multi-step routine operations

An agent that processes invoices by reading each one, pulling fields, looking up vendor codes, and generating a journal entry, can save a finance team hundreds of hours a month. This works because each step is well-defined and reversible. If the agent makes a mistake, a human catches it in the weekly close, and the cost of the error is bounded.

Coordinating other software

An agent that orchestrates several APIs to accomplish a task that previously required custom code can dramatically reduce engineering work. "Take this customer's address, check our shipping rates, generate a label, schedule a pickup" is the kind of work where an agent shines, provided the agent's actions are logged and reversible.

The pattern across all four is the same. The work is well-defined. The boundary of the work is explicit. A human can verify the output. The cost of an error is bounded and recoverable. Take any of those four conditions away and the agent becomes a liability.

Where agents fail, and how

I have watched enough agent deployments fail that I can now describe the failure modes by name.

The competent-confidence problem

Agents do not know when they are out of their depth. They will confidently take an action that would make any human pause. They will book the wrong flight. They will send the wrong email. They will make the wrong refund. They are not being malicious, they are doing what they always do, which is generating the most plausible next action given the situation. The plausibility of the action and the correctness of the action are two different things, and the agent cannot distinguish between them.

The cascading-failure problem

Agents operate in loops. Each step depends on the previous step. If step three is slightly wrong, step four is more wrong, step five is more wrong than that, and by step ten the agent is doing something that no human watching at step one would have predicted. There is no equivalent of a code review at step five. The agent just keeps going.

The blind-eye problem

When humans review every action an agent takes, the agent is useful. When humans skip the review because the agent has been right ninety-eight percent of the time, the two percent failure rate compounds quietly into a serious problem. The pattern I see most often is heavy supervision in the first month, then drift toward rubber-stamp review, then a major incident, then a tightening that lasts six weeks, then drift again. Plan for this. Build the supervision into the system itself, not into the discipline of humans who will, over time, get tired.

The accountability problem

When an agent makes a mistake that costs you a customer, you cannot fire the agent. You cannot promote the agent. You cannot have a development conversation with the agent. The agent has no career, no reputation, no incentive to learn. All the accountability flows back to the human who deployed it. This matters more than it sounds. It means that the responsibility for the agent's mistakes is yours, in a way that the responsibility for a junior employee's mistakes is shared.

Why you cannot fire your data scientist

This is the section that, if you skip the rest of the chapter, I most need you to read.

In the next twelve months, a vendor will sell you on an agent that, in a demo, will do something your data scientist does. The demo will be impressive. The vendor will, gently or not so gently, suggest that you could now do this work without the data scientist. The math will look good in a spreadsheet. The board will be excited.

If you act on this, you will discover the same thing I discovered when I came back from Italy. The work the data scientist does that the agent can replicate is the easy part. The work the data scientist does that the agent cannot replicate, which is most of the work, includes choosing which problems are worth solving, knowing when a model is starting to drift in production, recognising when a vendor is promising things the technology cannot deliver, building the relationships with the business teams that make the work usable, and noticing the subtle thing that says this analysis is not quite right and we should look again before shipping.

All of that is the equivalent of the experience my Stanford graduate did not have. The agent does not have it either. The agent will, just like she did, confidently do the parts of the job it can see and silently fail at the parts of the job it cannot. And just like with her, you will not know what was missed until it is too late.

> "Hire the agent. Do not fire the senior. The agent is a junior team member with very fast hands. It needs a senior nearby, the way every junior does, and for the same reasons."

The right pattern is to use the agent to give your data scientist leverage. Let the agent do the boring fast work. Let the senior person do the judgement work and oversee the agent's output. Your team gets meaningfully more productive. Your senior people stay. Your institutional knowledge stays. Six months later, when something subtle goes wrong, there is a competent human in the room who can spot it and fix it.

This is the boring, correct answer that vendors will not give you, because they sell more agents if you believe the agents replace people. It is the answer I would give you if you were paying me directly, and it is the answer I am giving you here.

A composite case

A consumer fintech company I advised in 2025 wanted to use agents to handle customer support. They were drowning. Ticket volume was up sixty percent year over year, hiring was capped, and customers were waiting too long for responses. The CEO had read a vendor pitch about agentic customer support and was, in his words, ready to write the cheque.

I asked him to wait a quarter and run a pilot. He did, somewhat reluctantly. Here is what happened.

They started with the narrowest possible deployment. The agent handled exactly one type of ticket, password resets, which made up about eighteen percent of the volume. The agent's job was to verify the customer's identity using their existing flow, reset the password, send a confirmation, and close the ticket. Every action was logged. The first hundred resolutions were reviewed by a human before any others were allowed. Then the next thousand. Then random samples of one in twenty, indefinitely.

In the first month the agent handled password resets with a 96 percent success rate. The 4 percent that went wrong went wrong in interesting ways. One customer was actually a fraudster who had compromised the legitimate account holder's email; the agent reset the password to the fraudster's specifications without noticing the warning signs a human agent would have caught. Two customers were children using their parents' accounts, which surfaced a compliance problem nobody had thought about. A handful of edge cases involved password reset attempts during account closure flows that the agent did not know existed.

The team learned. They added checks. They tightened the boundary of what the agent was allowed to do. By month three the agent was handling password resets reliably and they expanded the agent to a second ticket type, the next-most-common one. They repeated the pattern. Narrow deployment, intense supervision, expand only after the previous expansion is stable.

A year in, the agent handles roughly forty percent of incoming tickets across six narrow ticket types. The other sixty percent goes to humans. Customer satisfaction scores have gone up because routine tickets are resolved faster. The support team has not been laid off; it has been redeployed to handle the harder tickets that require judgement, with each agent spending more time per ticket because they are no longer drowning in password resets.

Most importantly, when the agent started to drift in month seven, because a base model update changed how it interpreted a particular instruction, the humans noticed within a day because they were still spot-checking. They paused the agent, fixed the instruction, resumed. If they had laid off the support team in month three because the agent looked so promising, no one would have noticed the drift, and the company would have spent a month sending wrong responses to thousands of customers before anyone realised.

The CEO has, on more than one occasion since, told me that the smartest thing he did with this project was wait the quarter.

Objections you will hear

"This vendor's demo showed the agent doing everything"

Demos are designed by the vendor to look impressive in a controlled setting. The real test is not the demo. The real test is a four-week pilot on your actual data with your actual workflows, with honest measurement of what worked, what failed, and what the failures cost. Insist on this. Vendors who refuse to run an honest pilot are usually vendors whose product does not work outside the demo.

"If we do not deploy agents aggressively, our competitors will pass us"

Possibly. They are also possibly about to spend significant money on premature agent deployments that will set them back two years. The right move is not the fastest deployment. It is the deployment with the highest sustainable ratio of value to risk. Run the pilots. Move where the math works. Pass on the parts where the math does not.

"The agent saves us so much money. The math is obvious."

The math is obvious if the agent works. The math is brutal if the agent fails in a way you do not catch. Build the failure scenario into the math. Ask yourself, if this agent makes the wrong decision a hundred times a month for six months before we notice, what is the cost? If that cost is small and recoverable, the math is fine. If it is large, your math is missing the most important number.

"The agent has been right ninety-eight percent of the time. We can trust it."

Ninety-eight percent right is two percent wrong. At the volumes most companies process, two percent wrong is a lot of customers having a bad experience. The right question is not the success rate. It is what happens to the customers who fall into the failure mode, and how you find out about them. A high success rate combined with no detection of failures is the most dangerous configuration there is.

"We will save by replacing senior staff with agents"

You will not save. You will create a six-month productivity bump followed by a multi-year recovery. The senior staff carry the judgement that the agents do not have. When something subtle breaks, and something subtle always breaks, the senior staff are who notice and fix it. Replace them and you have no one watching the agents. This is the most expensive false economy in the AI conversation right now. Do not make it.

Questions for your leadership team

36. For every agent we are considering deploying, what is the explicit boundary of what the agent is allowed to do? What happens when something falls outside that boundary?
37. How will we detect when the agent has started to drift or fail silently? Who watches the agent? How often? What is the trigger for tightening?
38. What does it cost us if the agent is wrong a hundred times a month for six months before we notice? Is that cost acceptable?
39. Have we decided whether any of our senior staff will be laid off because of this agent? If yes, are we sure we are not about to fire the person who would have caught the agent's mistakes?

40. Have we run a pilot on our actual data with our actual workflows, or are we deciding based on the vendor's demo?
41. If the agent is offline tomorrow because the vendor's model changed, can our business operate?

SIX THINGS TO TAKE FROM THIS CHAPTER

An agent is an LLM with hands. The hands are real. The judgement behind them is not.

The Stanford-graduate problem is the right frame. Impressive in narrow ways, dangerously confident in the rest. Treat agents accordingly.

Deploy agents narrowly. Define the boundary. Expand only after the current scope is stable.

Build the supervision into the system, not into the discipline of humans who will eventually skim.

Never fire your senior people on the strength of an agent demo. The senior people are who will catch the agent when it goes wrong.

The four conditions for a safe agent deployment: well-defined work, explicit boundary, human verification, bounded cost of error. Take any away and you have a liability.

ONE-LINE TAKEAWAY

Agents are not a miracle. They are a junior team member with very fast hands and no judgement. Hire them like you would hire a brilliant fresh graduate. Give them narrow work. Watch closely. Keep the seniors. Never go on the operational vacation in Italy that ends with you twenty customers down.

CHAPTER 7

The Generative AI Stack

In 2019 I drew a diagram for the CTO of a Fortune 500 company.

He had asked me what a typical data-science deployment looked like at his organisation. I drew three boxes on the whiteboard. A data warehouse. A model. An API. Connected them with three lines. That was the whole stack for most of what we did at the time.

Five years later I sat in the same conference room with his successor. She showed me the architecture for their first generative AI product. There were forty-three boxes. There were a dozen vendors. There was a vector database I had never heard of, an orchestration framework that had not existed three years before, and three different observability tools that were monitoring three different things that I could not, at first glance, distinguish from each other.

The thing being built was a customer-support assistant. The thing being shown to me was the modern AI stack.

She asked me, with some embarrassment, whether all of this was necessary. The honest answer is mostly yes. The other honest answer is that knowing why each piece exists, what it does, and which ones you actually need is now one of the most important things a technical leader can understand. So this chapter walks through it. Not exhaustively, because the specific tool names change every quarter. But by category, because the categories will hold.

What the stack used to be

Before generative AI, the stack for a typical machine-learning product had three real pieces. A place to store the data. A way to train the model. A way to serve the model at runtime. Everything else was supporting infrastructure that an experienced team could build in weeks.

The data lived in a warehouse. The model lived in a model registry. The serving layer was usually a small piece of code wrapped around the model, behind an API. Monitoring meant watching the API's latency and error rate. If you wanted to know whether the model was still accurate, you ran an offline evaluation once a quarter.

This world had its problems but it had a coherent shape. A reasonably senior engineer could draw the architecture on a napkin and the napkin would make sense for years.

Generative AI broke the napkin.

What is actually new

When the model in your system is an LLM rather than a custom classifier, several things change at once. The change is not just one new box on the diagram. It is a dozen new boxes, each of which exists because of a real problem the old architecture did not have to solve.

The new pieces, in the order they usually get added as a project matures.

Embeddings, and the place to store them

An embedding is a way of turning a piece of text into a list of numbers, in such a way that pieces of text with similar meaning end up with similar numbers. Roughly speaking, an embedding is the meaning of the text, expressed as coordinates in a high-dimensional space.

This sounds abstract. It is also the single most useful piece of new infrastructure in the AI stack, because it lets you do something we could not do well before: search by meaning rather than by keyword. "Find me the most relevant section of our employee handbook for this customer's question." Old-style search would look for matching words. Embedding-based search looks for matching meaning, even when the words are different. A customer asks about "time off" and the embedding finds the section about "PTO" without ever having seen the connection spelled out.

To make this work at scale, you need a place to store millions of embeddings and look up similar ones quickly. That place is called a vector database. The main ones in 2026 are Pinecone, Weaviate, Qdrant, Milvus, and a feature called pgvector that lets you do the same thing inside Postgres without adding a new database. They are not all equivalent. Pinecone is the easiest to start with. pgvector is the easiest to integrate into an existing stack. The others sit somewhere between.

For most teams just starting, pgvector is the right answer. You probably already have Postgres. You probably do not need millions of embeddings yet. Adding a new vendor for a problem your existing database can solve is a common over-engineering mistake. Reach for a dedicated vector database when you have measured that pgvector is no longer fast enough, not before.

Retrieval, the discipline of getting the right context

Vector search returns documents that are semantically close to the query. The next problem is what to do with those documents. How many to retrieve. How to rank them. How to assemble them into the prompt for the LLM. Whether to re-rank with a smaller model after the initial retrieval. Whether to chunk documents finely (better recall, more noise) or coarsely (less noise, worse recall).

These choices, collectively called retrieval engineering, are where most RAG systems go wrong. The chapter on LLMs talked about RAG as a pattern. This is where the engineering actually lives. A team that thinks retrieval is just "call the vector database" will ship a system that mostly works and occasionally produces wildly wrong answers, and they will not understand why. A team that treats retrieval as a discipline, with explicit chunking strategies, hybrid search combining keywords and embeddings, and a re-ranking step, will ship a system that is meaningfully more accurate.

This is one of the places I most often see leaders under-invest. The vector database is the cheap part. The retrieval engineering is the expensive part. The companies that get this right are not the ones with the best model. They are the ones with the best retrieval.

Orchestration

A real LLM application is rarely just one call to one model. It is usually several calls: one to interpret the user's request, one to retrieve relevant context, one to draft a response, one to check the response for safety, sometimes several more for tool use or sub-tasks. Coordinating these calls, handling failures, retrying when something times out, parallelising what can be parallelised, all of this is the work of an orchestration framework.

The popular ones in 2026 are LangChain and LlamaIndex on the open-source side, and a handful of commercial offerings that wrap similar functionality. Some teams use them happily. Other teams write their own orchestration in plain code and avoid the frameworks entirely, on the grounds that the frameworks add abstraction without solving the hard problems.

My honest opinion, having watched dozens of teams make this choice, is that the frameworks are useful for the first version of a project and a liability by the third version. They make it easy to glue components together quickly, which is great when you are exploring. They make it hard to debug subtle issues, which is bad when you are scaling. Most mature teams I know have ended up replacing the framework with hand-written orchestration after the first year. Plan for that path. Use the framework to learn the shape of the problem, then expect to replace it.

Prompt management

The prompts you send to your LLM are now part of your codebase. They change. They get versioned. They have bugs. They have regressions. Treating them as throwaway strings is a mistake that catches up with every team that makes it.

Mature teams treat prompts as first-class engineering artefacts. They live in version control, not in a database somewhere. Changes to prompts go through code review. Different prompts are A/B tested in production. There is a documented history of why each significant prompt change was made.

This sounds like overkill until your model behaviour mysteriously changes one Tuesday and nobody can figure out why. Then you discover that someone edited a prompt directly in the admin panel six weeks ago, no one wrote down what they changed, and your customer success team has been quietly fielding complaints ever since. I have seen this exact pattern at multiple companies. Build the discipline early.

Evaluation

How do you know if your LLM application is good? In the old world of classifier models, you ran the model against a held-out test set, measured accuracy, and got a number. With LLMs, the answer is much harder. The output is open-ended. There is no single right answer. "Good" depends on tone, helpfulness, accuracy, safety, format, and a dozen other things that resist a single metric.

Evaluation harnesses are the new discipline of measuring this. The basic pattern is to create a curated set of test inputs, run your application against them, and either have humans rate the outputs or have another LLM rate them according to a rubric you define. You track these ratings over time. When you change the prompt or the model or the retrieval, you re-run the evaluation and check whether you improved or regressed.

Most teams skip this step early in a project and pay for it later. The team that has an evaluation harness can iterate confidently because they can measure their progress. The team that does not has to guess whether each change made things better or worse, and the guesses are often wrong. Build the harness early. It will be slightly painful for two weeks and then save you months of debugging over the following year.

Observability

When something goes wrong in production with an LLM application, you need to know what happened. Which prompt was used. What context was retrieved. What the model returned. How long each step took. How much each call cost. Whether the safety filter triggered.

This is the work of LLM observability tools. Names in 2026 include Langfuse, Helicone, Arize, and a handful of others. The market is unsettled and consolidation is likely. The category is essential. You cannot debug what you cannot see, and LLM applications generate a kind of data your traditional logging tools were not designed for.

At minimum, you need to be logging every prompt sent, every response received, every error, every retrieval, and the cost of each call. You need to be able to search these logs by user, by session, and by output quality. The cost line is more important than it sounds. LLM costs can run away from you fast, and the only protection is being able to see, in real time, who is consuming how much.

Safety and content moderation

Your LLM will sometimes produce output you do not want shipped. Confidential information that leaked through retrieval. Personally identifying details. Inappropriate language. Content that violates your terms of service or the platform's policies.

Content moderation in the AI stack is usually a combination of filters from the model provider, custom rules you write for your domain, and another LLM acting as a checker. The right combination depends on the risk profile of your application. A customer support chatbot needs less protection than a medical advice tool, which needs less than a system that recommends financial actions. Match the strength of moderation to the stakes.

What you actually need versus what vendors will sell you

If you read the previous sections and concluded that every LLM project needs all of these components, vendors will love you. They will also bankrupt you, both in money and in operational complexity.

The honest minimum for most projects is much smaller than the diagram suggests.

- A way to call the LLM (the model provider's SDK).
- A way to manage prompts (a file in your code repository is fine for version one).
- A way to log what happened (your existing logging stack, plus a custom field for the prompt and response).
- A small evaluation harness (a Jupyter notebook with twenty test cases is fine for version one).
- Cost monitoring (a daily report from the model provider, until the volume justifies more).

That is the whole MVP stack. Everything else gets added when you measure that you actually need it, not on the assumption that you will. The forty-three-box architecture in the CTO's office is what mature deployments look like after years of evolution. It is not where you start.

> "Every box on your architecture diagram should exist because of a problem you have measured, not because a vendor told you it solves a problem you might one day have."

Build, buy, or rent each layer

For each piece of the stack, you have three options. Build it yourself in code. Buy a dedicated tool. Rent it as a service. Different layers have different right answers.

For the model itself, almost everyone should rent. Buying or building a model is a multi-million-dollar undertaking that almost no use case justifies. Use the frontier models when capability is paramount, the open-weight models when control matters more, and small specialised models when cost dominates.

For vector storage, build first using pgvector in your existing database, buy a dedicated vector DB only when you measure that you have outgrown it. Renting in this layer is almost never the right answer.

For orchestration, build first in plain code, then consider an open-source framework only if your patterns become repetitive enough to justify it. The commercial orchestration vendors are mostly solving problems that go away when you write the code yourself.

For prompt management, build first using files in your code repository, then move to a dedicated tool only when multiple teams need to collaborate on prompts.

For evaluation, build first using a simple harness in your existing test framework, then move to a commercial tool only when you have outgrown what you can maintain yourself.

For observability, this is the one layer where I would buy or rent earlier than the others. LLM observability is genuinely harder than traditional observability and the dedicated tools are meaningfully better than what most teams will build themselves. Worth the money.

Notice the pattern. Build first, buy or rent later when you have measured the need. The vendors will tell you the opposite. The vendors are wrong, except in observability, where they are mostly right.

A composite case

A mid-sized B2B software company I worked with built their first LLM-powered feature in 2024, a documentation assistant that answered customer questions about their product by retrieving from their own docs.

The first version, shipped in six weeks, was a single source file. It called the LLM with a long prompt that included the user's question and the most relevant section of the documentation, found by a simple keyword search. No vector database. No orchestration framework. No evaluation harness. The team had argued for all of these. The engineering lead had said no, ship version one first.

Version one worked, sort of. The answers were correct about seventy percent of the time. Customers liked it. The team started getting requests for more capabilities. They were also starting to see the failure modes more clearly. Keyword search was missing relevant sections that did not share words with the question. Some answers were confidently wrong. Cost was higher than projected because long prompts cost more.

Version two, three months later, added embedding-based retrieval using pgvector. Accuracy on the same test set went from seventy percent to eighty-six percent. The cost per query dropped because the prompts got shorter. The team also added their first evaluation harness, fifty test questions with hand-written ideal answers, run automatically against every change. The harness immediately caught a regression they would otherwise have missed.

Version three, six months after that, added a dedicated observability tool because the volume had grown enough that debugging customer complaints by reading log files was becoming painful. They added a simple prompt management system, a YAML file in the repo, so non-engineers on the documentation team could propose prompt changes without having to write code.

Version four, a year in, looked nothing like version one. It had nine identifiable components: model provider, pgvector, retrieval engineering layer, custom orchestration code, prompt YAML, evaluation harness, observability tool, cost monitoring dashboard, and safety filter. Every one of those nine had been added because the previous version exposed the need for it. None of them had been added because the team thought it would be cool to add.

The engineering lead told me later that the most valuable decision the team made was the discipline of measuring before adding. Every new component had to justify itself with data from the previous version. This is the opposite of how most teams build, which is to plan the complete architecture up front and then build all of it before shipping anything. The disciplined approach took the same total time and produced a much better system.

Objections you will hear

"We need to build the full architecture before shipping anything"

No, you do not. Ship the simplest thing that solves a real problem. Then measure where it falls short. Then add the next piece. The full architecture is what you have at version five, not what you start with at version one. Teams that try to build the whole stack before shipping usually do not ship.

"This vendor's all-in-one platform will save us months"

Sometimes true, often not. The all-in-one platforms work well when your use case fits exactly their mental model. They become painful when your use case drifts, because every component is coupled to every other. Build the simple version yourself first. Then evaluate the all-in-one platforms against what you actually need, not against the vendor's marketing.

"We are too late, our competitors are ahead"

Probably not. The companies that look ahead because they shipped early often have technical debt they will spend the next two years repaying. The companies that look behind are sometimes about to ship the right thing while the early movers are debugging the wrong thing. Speed is overrated in this category. Correctness is undervalued.

Questions for your leadership team

42. For each layer of our AI stack, have we built, bought, or rented based on measured need or on assumption?
43. Do we have an evaluation harness? If not, how do we know when we have improved the system?
44. Are we logging every prompt, response, retrieval, and cost? Can we search this for any given customer complaint?
45. Are our prompts in version control? Or could anyone change them in production without a paper trail?
46. If our model provider doubled their prices tomorrow, how long would it take us to switch?
47. What is the simplest version of this product we could ship in six weeks? Have we considered shipping it?

SIX THINGS TO TAKE FROM THIS CHAPTER

The modern AI stack has more boxes than the old one. Most of them are real. Few of them are needed on day one.

Embeddings let you search by meaning, not keywords. The vector database is cheap. The retrieval engineering around it is expensive.

Build first, buy or rent only when you have measured the need. Vendors will sell you the opposite. The exception is observability, where buy is usually right.

Prompts are code. Version them. Review them. Test them. Do not edit them in production admin panels.

Evaluation harnesses are the discipline that lets a team improve confidently. Teams without them are guessing.

The first version of an LLM product should fit in one source

> file. The forty-third version will not. Build between the two by adding only what you have measured a need for.

ONE-LINE TAKEAWAY

The generative AI stack is real and the components are necessary at scale. They are not all necessary on day one. Build the smallest version that solves a real problem. Add components as the data tells you to. Treat the architecture as something that evolves with your understanding of the problem, not something you design once and ship complete.

PART III
The Build

Designing systems that work, securing them, and the teams that run them.

CHAPTER 8

Architecting AI That Doesn't Break

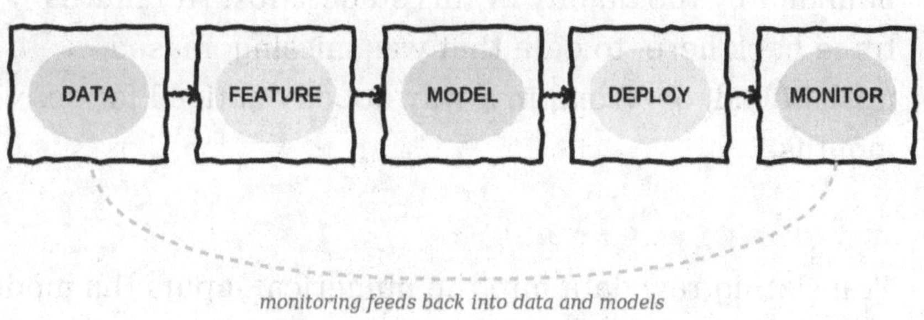

monitoring feeds back into data and models

Most AI projects fail. The failure rate, depending on which industry analyst you trust, is somewhere between sixty and eighty percent. The failures rarely happen because the model didn't work. They happen because the model worked in a notebook and never made it into production, or made it into production and quietly drifted into uselessness without anyone noticing.

Good AI architecture is the discipline of building systems that survive contact with the real world. This chapter is the high-level map. Engineering teams will implement the details. Leaders need to know the shape, so they can ask the right questions in design reviews.

The five stages of an AI system

Every production AI system has the same five stages, even when teams pretend otherwise.

1. Data

Where the inputs come from. Customer transactions, sensor readings, support tickets, photos, voice recordings. The quality of everything downstream is bounded by the quality of this stage. Most AI failures trace back here, to data that was missing, biased, mislabelled, or wrong in a way nobody noticed for six months.

2. Feature engineering

Translating raw data into the numerical inputs the model actually uses. Sometimes trivial (the customer's age, the timestamp). Sometimes the entire game (which combinations of transactions look like fraud). Modern deep learning has reduced the need for hand-crafted features in some domains. It has not eliminated it.

3. Model

The mathematical machine that takes features and produces predictions. Where most public attention focuses. Often, in practice, the least interesting part of the system, because off-the-shelf models are now extremely good.

4. Deployment

Getting the model out of a notebook and into a system that real users or systems can hit. This is where most AI projects die. The journey from research code to production code is harder than it looks. Often more expensive than the research itself.

5. Monitoring

Watching the deployed model for accuracy degradation, data drift, unusual usage patterns, and emerging failure modes. Most teams skip this. Most teams find out about model failures from angry customers.

> "A model in production is a system that needs constant supervision. A model in a notebook is a hobby."

Where projects die: the deployment cliff

The biggest gap in most AI organisations is between the data scientist who built the model and the engineering team that needs to run it in production. The two groups often have different skills, different tools, and different incentives. Bridging this gap is the single most important infrastructure investment a leader can make.

In the field, this discipline now has a name: MLOps. Machine learning operations. It borrows heavily from DevOps but adds AI-specific concerns, model versioning, data versioning, monitoring for distribution shift, automated retraining triggers, and explainability tooling.

Three practical investments make the cliff smaller.

48. A feature store. A central place where engineered features are stored, versioned, and reused. Solves the problem of the model in production seeing slightly different features than the model in training.
49. A model registry. A central place where trained models are stored with their version, metrics, and metadata. Solves the problem of "which model is actually running in production right now?"
50. Automated monitoring. Dashboards and alerts that track prediction distributions, input data drift, and accuracy on a holdout set, automatically, with humans paged when thresholds break.

None of these are exciting. All of them quietly determine whether your AI investment is alive or dead in eighteen months.

Designing for scale, plainly

Most leaders worry about scale too early and the wrong kind of scale. Two principles.

Scale to your real load, with headroom, not for a hypothetical future

Most AI systems do not need to serve a billion users on day one. Build for the load you actually have, with a clean architecture that can grow. The most expensive engineering decisions I have seen were premature optimisations for scale that never materialised. The most expensive operational mistakes were systems that grew faster than expected and had to be re-architected mid-flight. Both are real risks. Honest forecasting and modular design beat both.

Cost scales differently than load

Cloud costs do not scale linearly. A model that costs $100 a month to serve at 1,000 requests per day may cost $50,000 a month at 1,000,000. Sometimes this is fine and the revenue scales accordingly. Sometimes it is fatal. Run cost models alongside load models. Decide which workloads must be cheap and which can be expensive. Architect accordingly.

A composite case: a healthcare provider's first AI system

Here is a composite drawn from several real projects I have advised on. No specific company; all the details are common patterns.

A large healthcare provider wanted to use AI to predict which patients were at high risk of readmission within 30 days of discharge. The clinical case was clear: a follow-up call from a nurse to high-risk patients reduces readmission by a meaningful margin. The problem was identifying which patients needed the call.

The first version of the model worked beautifully in the data scientist's notebook. Trained on three years of historical patient data, it predicted 30-day readmission with reasonable accuracy. The team celebrated, scheduled a launch, and discovered that almost nothing was ready for production.

The data the model needed had to be assembled from four separate clinical systems on different schedules. The discharge summary the model relied on was sometimes written hours after the patient left, by which point the prediction was useless. The model's confidence scores meant nothing to the nurses who would receive them, so a calibration layer had to be added. Patient demographics in the training data did not match the demographics of the patient population the model would actually see, so the predictions skewed in ways that required correction. The audit trail required by hospital regulators had not been considered.

Six months later, the model was in production. It worked. It quietly improved readmission rates by a meaningful margin and continues to do so. The lessons were not about the model. They were about everything around the model, the data pipelines, the calibration, the deployment infrastructure, the monitoring, the governance. The model was maybe ten percent of the actual work.

Objections you will hear

"MLOps is too much overhead for our small team"

It feels like overhead the day you build it. It feels like the foundation of your business the first time a model fails in production at three in the morning. The teams that skip MLOps spend more total engineering time on AI than the teams that build it, because every model in production becomes a permanent debugging burden when there is no system to support it. Build it small at the start, but build it.

"We will add monitoring after we ship"

You will not. The cost of adding monitoring after a model has been in production for six months is much higher than building it in from the start, because you now have to instrument a system you have stopped thinking about and validate the instrumentation against behaviour you can no longer easily reproduce. Monitoring goes in before the model ships, or it usually does not go in at all.

"The model accuracy is what matters, not the infrastructure"

Model accuracy is one of perhaps fifteen things that decide whether your AI product succeeds. The other fourteen include deployment reliability, latency, cost per query, drift detection, calibration, audit trails, and the ability to roll back when something goes wrong. A 95-percent accurate model in a broken system performs worse than a 90-percent accurate model in a robust one.

"Our engineers can handle this without a framework"

Sometimes true for very small teams. Becomes false faster than people expect, usually when the second or third model goes into production and the team realises they are now rebuilding the same scaffolding from scratch each time. The framework does not have to be commercial. It has to be intentional.

Questions for your leadership team

51. For each model in production, do we know when we last measured its accuracy? Do we know if it has drifted?

52. Do we have a documented deployment process, or does each launch effectively reinvent the wheel?
53. Who is paged when a model fails in production? Do they know what to do? When did they last practice?
54. What is the lag between a model starting to drift and our detection of the drift?
55. If our most senior ML engineer left tomorrow, could a new hire deploy the next model within thirty days?

IF YOU REMEMBER NOTHING ELSE FROM THIS CHAPTER

Five stages: data, features, model, deployment, monitoring. Skip any of them and the system breaks.

Most AI projects die at the deployment cliff. Invest in MLOps as if your AI strategy depends on it. It does.

The model is usually ten percent of the work. Plan accordingly.

ONE-LINE TAKEAWAY

Architecting AI is the boring discipline of building systems that survive the real world. The model is the small part. The data pipelines, deployment, and monitoring are the work.

CHAPTER 9

Securing What You Build

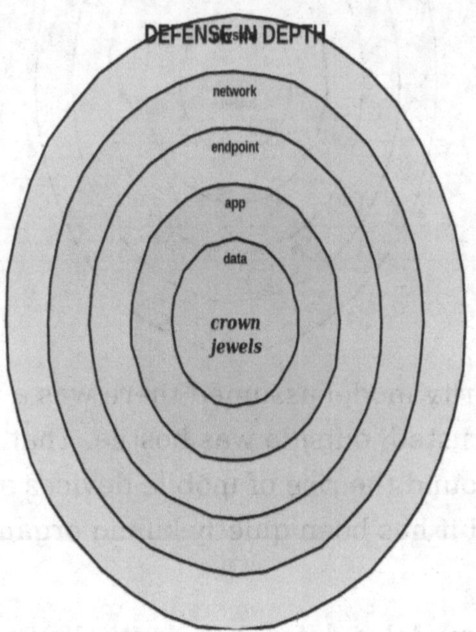

Cybersecurity used to be a back-office concern. It is now a leadership concern, because the cost of a serious breach can take a company down. AI and data systems make the surface area bigger. They also introduce a new class of attack that traditional security teams are not trained to handle.

This chapter is the leader's view. The detail belongs to security engineers. The framing belongs to you.

Defense in depth, the only model that actually works

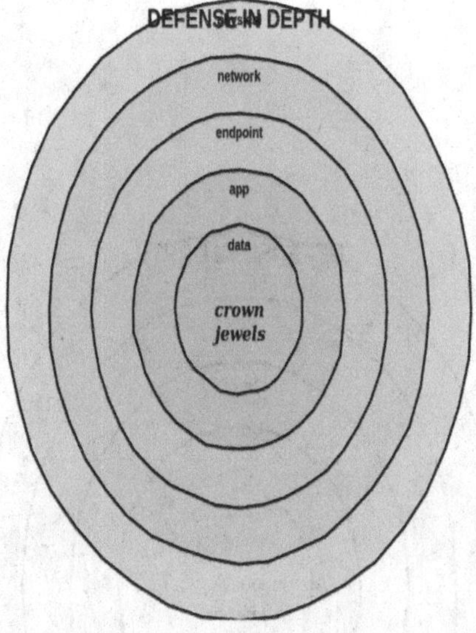

The old security model assumed there was a perimeter. Inside was trusted; outside was hostile. That model died sometime around the rise of mobile devices and cloud services, and it has been quietly killing organisations ever since.

The modern model is defense in depth. You assume that any single layer of security will eventually fail, and you build redundant layers so that a failure at any one does not compromise the whole system. Onion, not eggshell.

Five layers, from outside in: physical security (locked data centres, badge access), network security (firewalls, encrypted traffic, VPNs), endpoint security (laptops, phones, IoT devices), application security (authentication, authorisation, secure coding), and data security (encryption at rest, encryption in transit, access controls). Each layer fails differently. Each layer catches different threats. None of them is enough on its own.

Identity and access: the highest-leverage investment

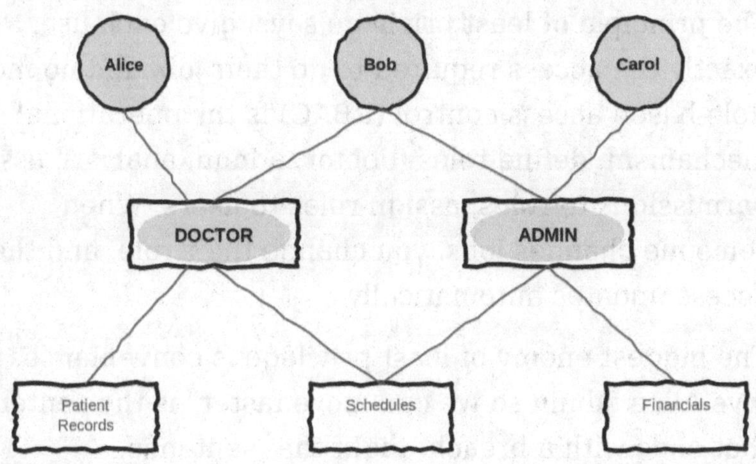

Most security incidents I have seen up close started with credential theft. Someone's password was phished. Someone's laptop was lost. An ex-employee's access was never revoked. The attacker did not break the firewall; they walked in through a door that had been left open.

Two practices, applied consistently, prevent most of these incidents.

Multi-factor authentication, everywhere

Passwords alone are not enough. Multi-factor authentication, where logging in requires both something you know (a password) and something you have (a phone, a security key), defeats almost all phishing and credential reuse attacks. MFA should be non-negotiable for every employee, especially for systems that touch production data.

Role-based access control, with least privilege

Most employees have access to far more than they need. The principle of least privilege says, give each user exactly the access required to do their job, and no more. Role-based access control (RBAC) is the operational mechanism: define roles (doctor, admin, analyst), assign permissions to roles, assign roles to users. When someone changes jobs, you change their role, and their access updates automatically.

The biggest enemy of least privilege is convenience. "Just give Alice admin so we can move faster" is the sentence that ends with a breach. Fight that sentence.

AI-specific threats: the new attack surface

AI systems introduce attacks that traditional security people are not yet equipped to think about. Three matter most.

Adversarial inputs

Carefully crafted inputs that fool the model. A few perturbed pixels on a stop sign can cause a vision system to read it as a speed limit sign. A subtly worded prompt can convince a language model to ignore its safety training. The defence is adversarial training (deliberately exposing the model to attack examples during training) and input validation (catching obviously suspicious inputs before they reach the model).

Data poisoning

Attackers contaminate the training data so the model learns the wrong patterns. Particularly dangerous in systems that continuously retrain on user-supplied data. The defence is rigorous data validation, anomaly detection on training data, and segregation of trusted and untrusted data sources.

Model extraction and inversion

Attackers send many queries to a deployed model and reconstruct either the model itself (theft) or sensitive properties of the training data (privacy leak). The defence includes rate limiting, output noise injection, and careful access controls on prediction endpoints.

None of these are theoretical. All of them have been demonstrated against production systems. Your security team should be reading the AI security literature. If they aren't, that's an organisational problem worth fixing now.

A composite case: a financial firm's security overhaul

A large financial services firm I worked with had grown through acquisitions. Each acquired company brought its own systems, its own access patterns, its own credentials. The result, fifteen years in, was a security posture that nobody fully understood, with overlapping access grants, dormant accounts, and inconsistent encryption.

The overhaul took eighteen months. Highlights:

- A complete inventory of every system and every account with access to it. This alone took three months. It revealed roughly twenty percent more accounts than the firm believed it had.
- Mandatory multi-factor authentication for every employee, with no exceptions. Two months of pain. Then forgotten.
- A single identity provider feeding access decisions to every system. Killed the patchwork.
- Role-based access control rebuilt from scratch. Every user reassigned to a small number of well-defined roles. The number of unique permission grants dropped by roughly ninety percent.
- Continuous monitoring with automated alerts on unusual access patterns. Two genuine credential-theft incidents caught in the first six months that would not have been caught before.

The cost was significant. The result, eighteen months later, was a security posture the chief information security officer could actually describe in a single afternoon, and an incident response time measured in hours instead of weeks.

Objections you will hear

"We have a CISO, they handle security"

Your CISO probably handles security excellently for the categories they were trained on. AI introduces categories most CISOs have not yet trained on. Adversarial inputs, prompt injection, model extraction, training data poisoning, these are real attack vectors that traditional security playbooks do not yet cover. Your CISO needs to be reading the AI security literature, not just relying on the security frameworks of 2019. If they are not, that is the gap to close.

"AI security is just regular security"

The basics overlap. Defense in depth, MFA, least privilege, all of that applies. The differences are real. Adversarial inputs attack the model itself, not the surrounding system. Data poisoning attacks the training pipeline. Model extraction attacks the deployed model through query patterns. These have no equivalent in traditional security and require their own thinking.

"We have never been attacked, we will be fine"

Most organisations who think they have not been attacked have been attacked and have not detected it. AI systems are an attractive target because they sit at the intersection of valuable data and immature defences. The probability that you will be attacked in the next two years is high. The probability that you will detect the attack is much lower. Build defences as if the attack has already happened, because it usually has.

"Adversarial attacks are theoretical"

They were theoretical in 2018. They are operational in 2026. Adversarial inputs have been used in real attacks against vision systems, content filters, fraud detectors, and authentication systems. The research is now applied. Your model is a target. The question is whether your defences keep up with the attacks, not whether the attacks exist.

"We cannot afford the security overhead"

You can afford it less than you think. The full cost of building defence in depth for AI is meaningfully smaller than the cost of a single serious breach. The 2024 average cost of a healthcare data breach in the US was around ten million dollars per incident. The cost of a competent AI security program at a similar company is a fraction of that, even at the high end. The math is on the side of investing early.

Questions for your leadership team

56. When did our security team last train specifically on AI-era threats like prompt injection and adversarial inputs?
57. Do we know our exposure to prompt injection across our deployed LLM applications? Have we tested for it?
58. What is our incident response plan if our AI model is compromised or extracted? When did we last rehearse it?

59. Have we audited which third parties (vendors, contractors, partners) have access to our AI systems and the data they touch?
60. If our model weights were stolen tomorrow, would we know? What would the business impact be?

IF YOU REMEMBER NOTHING ELSE FROM THIS CHAPTER

Defense in depth is the only model that survives contact with reality. Build layers; assume each will eventually fail.

Multi-factor authentication and least-privilege access are the two highest-leverage investments. Most breaches happen where these are missing.

AI introduces a new attack surface. Adversarial inputs, data poisoning, model extraction. Your security team needs to learn this.

ONE-LINE TAKEAWAY

Security is no longer optional, and AI systems expand the attack surface in ways traditional security doesn't yet cover. Defense in depth, MFA everywhere, least privilege, and a security team that reads AI papers.

CHAPTER 10

The Team Behind the Tech

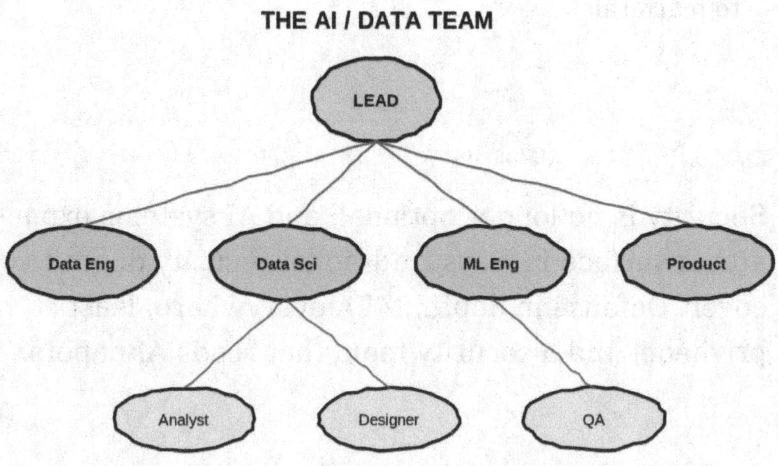

Every AI strategy I have seen succeed had a strong team behind it. Every AI strategy I have seen fail had either no team or the wrong team. The technology decisions matter, but they are downstream of the hiring and management decisions.

This chapter is about how to build and lead a data and AI team that does work the organisation actually uses.

Five roles, what each one actually does

Data engineer

Builds the pipes that move data from where it is generated to where it can be used. SQL, distributed systems, infrastructure. Often the most underappreciated role on the team. The first person you should hire, almost always.

Data scientist

Analyses data to find patterns, builds and trains models, runs experiments. The role with the loudest job title and the most variable skill base. Hire carefully. Ask for portfolios of actual work.

Machine learning engineer

Takes models out of notebooks and into production. Engineering depth combined with model fluency. The rarest and most expensive role on the market in 2026. Often a combination of data scientist and software engineer rather than a separate breed.

Product manager (for AI products)

Owns the connection between user need and what the team builds. AI product management is a specific skill, not generic product management with AI buzzwords. The best AI PMs understand uncertainty, can interpret model metrics, and know when to ship and when to wait.

Analyst

Translates business questions into data queries and back. Sometimes overlapping with data scientist, sometimes a distinct role. Often the team's connection to the rest of the business.

Most teams do not need all five from day one. Most teams should hire in this rough order: analyst (if you don't have one already), data engineer, then split based on whether your work needs more analysis (data scientist) or more production engineering (ML engineer).

How to actually hire well

Three principles that have served me well across hundreds of interviews.

Hire for what they have built, not what they have studied

Degrees correlate weakly with practical skill in data and AI. Portfolios correlate strongly. Ask candidates to show you something they built end to end, ideally something messy with real-world constraints. Listen for what they learned from the messy parts. Candidates who only talk about textbook techniques are early-career. Candidates who talk about the painful operational lessons are senior.

Hire for the ability to explain

The best data and AI people I have worked with can explain their work to a non-technical executive in three sentences. Without simplifying it to the point of being wrong. This is a rare skill. Test for it in interviews. Give them a complex topic from their resume and ask them to explain it to you, then ask them to explain it again as if you were a different audience.

Diversify deliberately

Teams that are demographically and intellectually diverse build better AI. This is not a moral claim, though it is also that. It is an empirical claim. Homogenous teams build models that work well for people like them and badly for people like everyone else. The bias issues that plague AI systems are downstream of who built them. Fix the team and you fix half the problem.

Leading the team: a few specific habits

Defend deep work

The best work data scientists and engineers do happens in long, uninterrupted blocks. Meetings, slack, status updates, ad-hoc requests all destroy this. The single most useful thing a leader can do for a technical team is to protect their calendar. No meetings before noon on two days a week is a start.

Make it safe to ship imperfect work

Perfectionism is the enemy of shipped systems. The best teams ship something workable, learn from it in production, and improve. The worst teams polish indefinitely and ship nothing. Reward the shipping. Forgive the imperfections.

Be present in design reviews

Leaders who skip technical design reviews end up with systems they don't understand and can't defend. You don't need to write the code. You do need to be in the room when the architectural decisions are made, asking what could go wrong, what we are betting on, and what we will do when we are wrong.

A composite case: the team that quietly transformed an insurer

A mid-sized insurance company built a data and AI team from scratch over three years. Lessons from how they did it.

They hired a data engineer first, before they hired a data scientist. This sounds backwards. It was correct. They spent the first six months building reliable pipelines and a clean data warehouse. The second six months, they hired two analysts to start producing reports that the business actually used. Only then did they hire a data scientist.

The order matters. The data scientist they eventually hired walked into clean pipes and engaged business users with real questions. Their work was immediately useful. Compare with the more common pattern of hiring a data scientist first, who then spends a year cleaning data and struggling to find willing internal customers. Most data scientists in that situation leave.

Three years in, the team had ten people, had shipped a dozen production models, and was responsible for measurable improvements in claims processing time, fraud detection, and customer retention. The chief data officer who built it told me the most important decision was the hiring order. The technology came easy after that.

Objections you will hear

"We can hire a data scientist first and build the rest of the team later"

The most common hiring mistake in AI. The data scientist arrives, discovers there is no clean data and no pipeline to clean it, and spends the next year doing data engineering work they did not sign up for. Most of those data scientists leave within eighteen months. Hire the data engineer first, almost always. The data scientist you hire next will land into a working pipeline and produce useful work in months instead of years.

"We cannot afford a full team, we will use contractors"

Contractors can be useful for specific bounded projects. They are bad substitutes for permanent team members on a long-term AI program. The institutional knowledge, the relationships with the business, the willingness to maintain systems years after they were built, all of this lives in permanent staff and does not transfer to contractors. The savings on the contractor are often less than the cost of the maintenance and rebuild work that follows.

"Engineering talent is too expensive in our market"

Sometimes locally true, almost always globally addressable. Senior AI talent can work remotely. Senior AI talent will accept lower base salary in exchange for genuinely interesting work. The companies that say they cannot afford the talent often have not done the work to find out what would actually attract it. The talent is more available than the market price suggests, if you are willing to compete on dimensions other than cash.

"We do not need diversity, we need capability"

This framing is wrong. The empirical question, repeatedly studied, is whether diverse teams build better AI systems. They do, particularly for systems that affect diverse user populations. The framing of "diversity versus capability" misses that diverse perspectives are themselves a capability. Teams that lack it ship products that fail in predictable ways for users who do not look like the team.

"Senior engineers do not want to work for us at our pay scale"

Possibly true, possibly addressable. If the answer is genuinely that you cannot pay competitive senior salaries, you have a strategic problem larger than this hiring decision. If the answer is that you can pay but have not, the fix is to pay. The cost of paying senior salaries is meaningfully less than the cost of running an AI program with insufficient senior talent.

Questions for your leadership team

61. Do we have a data engineer? If not, what is our data scientist actually spending their day doing?
62. When we interview AI candidates, do we ask for portfolios of actual work, or just resumes and degrees?
63. How diverse is our AI team, by every meaningful measure? When did we last look at the numbers honestly?
64. What is our retention rate on AI talent over the last two years? Is it where we want it to be?
65. Are we paying competitively for the level of AI talent we say we need, or are we underpaying and hoping?

IF YOU REMEMBER NOTHING ELSE FROM THIS CHAPTER

Hire a data engineer before a data scientist. Almost always.

Hire for portfolios and for the ability to explain. Degrees correlate weakly with practical skill.

Diverse teams build better AI. This is an empirical claim, not just a moral one.

The leader's job is to defend deep work, reward shipping over polishing, and stay present in design reviews.

ONE-LINE TAKEAWAY

The team decides whether your AI strategy works. Hire the data engineer first, hire for what they have built, diversify deliberately, and defend their calendar.

CHAPTER 11

Data Engineering, the Discipline

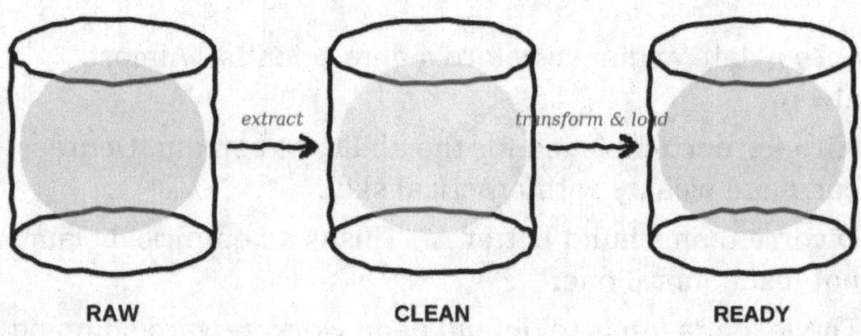

Data engineering is the least glamorous discipline in the AI world. It is also, by some distance, the one that determines whether anything else works.

This chapter is short because the principles are short. The execution is long, and the execution is where data engineers earn their salary.

What data engineering actually is

Data engineering is the practice of moving data from where it is generated to where it can be used, at the quality and speed the business needs. That is the whole job, expanded across thirty disciplines: ingestion, storage, transformation, validation, governance, lineage, security, performance optimisation, and a dozen more.

The discipline has evolved through three rough eras.

Era 1: The warehouse era (1990s, 2000s)

Data was extracted from transactional systems overnight, transformed by complex SQL pipelines, and loaded into a central data warehouse for analytics. Slow, structured, expensive. The dominant pattern for two decades.

Era 2: The big data era (2010s)

Hadoop, then Spark, allowed processing of data sets too large for traditional warehouses. The promise was that storage and compute would be cheap enough to keep everything. The reality was that the tools were brutal, the talent was rare, and most organisations ended up with expensive data lakes that the business could not use.

Era 3: The modern stack (2020s)

Cloud-native warehouses (Snowflake, BigQuery, Databricks) brought back the warehouse model with vastly better scale and economics. Tools like dbt made transformations maintainable. Streaming pipelines using Kafka and Flink made near-real-time processing realistic. The result is the most productive data engineering era we have ever had, and the one where leaders who don't keep up fall furthest behind.

ETL, ELT, and why the order matters

Two acronyms describe how data gets from source to usable form.

ETL: Extract, Transform, Load

The classical pattern. Extract data from source systems, transform it on the way (clean, aggregate, join), and load it into the warehouse already polished. Used to be the only option because warehouses were expensive and storage was scarce.

ELT: Extract, Load, Transform

The modern pattern. Extract data, load it into the warehouse as-is, transform it inside the warehouse using its scalable compute. Possible now because cloud warehouse storage is cheap and cloud compute is elastic. Faster, more flexible, easier to debug.

Most modern data stacks should default to ELT. The transformation step happens in SQL, in the warehouse, with proper version control through tools like dbt. The historical reasons for ETL are mostly gone. Some specific cases (regulated industries, very large volumes, real-time requirements) still need ETL or streaming. Most cases don't.

Data quality, the boring foundation

Models trained on bad data make bad predictions. Reports built on bad data make bad decisions. Most data quality work is invisible until it fails publicly, by which point recovery is expensive.

Three habits matter more than any tool.

Validate at ingestion

Every data source should have automated validation at the point it enters your system. Schema checks, range checks, freshness checks, volume checks. When validation fails, alert someone immediately. Most data quality disasters happen because nobody noticed the source feed had been broken for a week.

Track lineage

For every important data set, you should be able to answer: where did each column come from, what transformations were applied, who can change it, and when was it last updated? Modern tools (Atlan, Datahub, OpenLineage) do most of this automatically if you wire them in. Wire them in early.

Make data quality a metric

If quality isn't measured, it isn't managed. Define a small number of data quality metrics for your most important data sets. Report them weekly. Treat declines like you would treat declines in revenue or customer satisfaction.

A composite case: cleaning a decade of mess

A retail company I advised had a decade of analytics built on a dataset that everyone in the business quietly distrusted. Sales numbers from the warehouse did not match sales numbers from the website's analytics platform. Marketing attribution disagreed with finance reconciliation. Forecasts based on the data were routinely off by margins that everyone had stopped questioning out loud.

The fix took a year and was not glamorous.

The team rebuilt the core sales fact table from scratch, using the raw transaction log instead of derived snapshots. They added automated validation at every step. They documented every transformation, with the business owner who could explain why each one existed. They built dashboards showing data freshness, completeness, and quality scores. When discrepancies were found between systems, they were investigated and resolved, not papered over.

A year in, the business had a single sales number it trusted. The fights between marketing and finance about whose numbers were right stopped. Forecast accuracy improved measurably, not because the forecasting model changed but because the inputs were finally clean. The team's CEO told me later that the year of data cleanup was the single highest-return technology investment the company had made in a decade. There was no model. Just clean data.

Objections you will hear

"Data engineering is plumbing, the real value is in the models"

The plumbing decides whether the models work. A perfect model trained on bad data produces confidently wrong predictions. A modest model trained on clean data produces useful ones. The companies that have figured out AI at scale invariably invested heavily in data engineering first. The companies that skipped data engineering are the ones whose AI programs quietly produce nothing of value.

"We have a data warehouse, we are fine"

A data warehouse is one piece of the picture. Data engineering also includes ingestion, validation, transformation, lineage, quality monitoring, and governance. Most companies that say they "have a data warehouse" mean they have a place to put data, not that they have a working data engineering practice. Check the difference. The gap is usually larger than expected.

"Data quality issues are edge cases"

They are not edge cases. They are the dominant failure mode of analytics and AI work. Industry surveys consistently find that data scientists spend 60 to 80 percent of their time on data quality and preparation work. If that statistic surprises you, your team probably spends the same time and you have not been measuring it. The fix is not to ignore the quality work; it is to do it deliberately, with the right tools and people, so the data scientists can do data science.

"We do not have time to clean the data before shipping"

You will spend ten times as much time fixing the consequences of bad data after shipping. The math is always the same. The question is whether you prefer the cost up front, when you can plan for it, or under pressure, when you cannot.

"ETL has worked for us for twenty years, why change?"

Sometimes nothing needs to change. Sometimes ELT in a modern warehouse would give you ten times the flexibility and half the maintenance burden. Worth evaluating, not as a religious conversion but as an honest comparison. The teams that have moved tend not to go back, which is some evidence.

Questions for your leadership team

66. Can we trace any KPI on our executive dashboard back to the raw data that produced it, with documented lineage?
67. When was the last time two of our reports disagreed and we could not say with confidence which was right?
68. Do we have a measured data quality SLA, or do we discover problems when business users complain?
69. Do we validate data at ingestion, or do we discover problems downstream after they have spread?
70. Honestly, what fraction of our engineering and data science time goes to data quality issues each quarter? Do we know?

IF YOU REMEMBER NOTHING ELSE FROM THIS CHAPTER

Data engineering decides whether AI works. The model is downstream of the data.

Default to ELT in cloud warehouses. The historical reasons

for ETL are mostly gone.

Validate at ingestion, track lineage, measure quality. Bad data caught early is cheap; bad data caught late is expensive.

A year cleaning data is often the highest-return investment a company can make.

ONE-LINE TAKEAWAY

Data engineering is the discipline that decides whether everything else works. Default to ELT in a modern cloud warehouse, validate at ingestion, track lineage, and treat data quality as a measured metric.

PART IV

The Long View

Governance, ethics, economics, and how to stay useful as the ground keeps moving.

CHAPTER 12

The Letter You Do Not Want to Receive

There is a specific phone call that every leader who deploys AI now dreads. An increasing number have started getting it.

It comes from your legal counsel, or sometimes directly from a regulator. It says, in some variation, "We have some questions about the AI system you launched in March. We need to talk about how it was tested, what data it was trained on, how decisions are reviewed, and whether you have the documentation we are about to ask you for."

In 2023 this call was rare. In 2026 it is routine. By 2028 it will be expected. The shift, from AI as a wild frontier to AI as a regulated industry, has been the fastest regulatory transition of any major technology in my professional lifetime. And most leaders are still operating as though it has not happened.

This chapter is the catch-up. What changed. What it means in practice. What you need to build now, before the call arrives. None of this is legal advice. All of it is the framing that the leaders I respect most have already adopted.

The shift that happened, in three sentences

In 2024 the European Union passed the AI Act, the first comprehensive AI regulation in any major jurisdiction. In 2026 it began to be enforced, with phased deadlines that run through 2027. By 2028 most of the world's major markets will have some form of AI regulation, often modelled on the EU framework, often more restrictive in their own sectors.

Three sentences, but the implications take a chapter to unpack. The short version is that AI is no longer governed only by your own taste, your customers' patience, and the threat of bad press. It is governed by law. And the laws have teeth.

The EU AI Act, as one example, allows fines of up to seven percent of global annual turnover for the most serious violations. Seven percent of global annual turnover is not a regulatory inconvenience. It is the kind of number that gets a CFO's attention. It is also the kind of number that makes the decisions in this chapter no longer optional.

The EU AI Act, in plain language

I am not a lawyer. Get one. What I can give you is the conceptual frame the lawyers will assume you already understand when they sit down to advise you.

The Act sorts every AI system into one of four risk tiers, and the obligations on you depend on which tier your system falls into.

Tier 1: Unacceptable risk

Banned outright. Social scoring of citizens by governments. Real-time facial recognition in public spaces, with narrow exceptions. AI that manipulates human behaviour in ways that cause harm. AI that exploits vulnerabilities of specific groups. If your system is in this tier, you cannot deploy it in the EU. Most legitimate businesses are not in this tier, but some surprising things are. Read the list with your lawyers.

Tier 2: High risk

Permitted but heavily regulated. The category that catches most enterprise AI applications. This includes AI used in employment decisions (hiring, firing, promotion, performance), education (admissions, grading), credit scoring, law enforcement, migration, justice, and operation of critical infrastructure. Also AI that is a safety component of a regulated product, like an AI-powered medical device.

If your system is in this tier, you have a substantial compliance burden. You must maintain detailed technical documentation, conduct risk assessments, ensure human oversight, log every decision, monitor accuracy in production, and register the system in an EU database. You must do this before you deploy, not after. The fines for shipping a high-risk system without compliance are, as noted, ruinous.

Tier 3: Limited risk

Permitted with transparency obligations. Chatbots, AI-generated content, deepfakes. The main obligation here is to tell users they are interacting with AI, and to label AI-generated content as such. This is the tier most consumer applications fall into.

Tier 4: Minimal risk

Permitted with no specific obligations under the Act, though general consumer protection and data protection laws still apply. Things like spam filters, AI used to improve internal search, recommendation engines for non-essential consumer products. Most AI applications are in this tier and the Act will not change how you operate them.

The most important practical step is to figure out, for every AI system you have deployed or are about to deploy, which tier it lives in. This is not a one-time exercise. The same system can move between tiers as you change its use case. An assistant that helps employees write emails is Tier 4. The same model, repurposed to score employees for promotion, becomes Tier 2 overnight. Your inventory of AI systems and their classifications has to be a living document.

The United States patchwork

If the EU has produced a single coherent framework, the United States has produced fifty parallel conversations, and that is approximately the situation in 2026.

There is no federal AI law. The federal regulators that exist (FTC, EEOC, FDA, financial regulators) have started applying their existing authorities to AI within their domains. The FTC has gone after companies for deceptive AI marketing. The EEOC has gone after employers whose AI hiring tools showed bias. The FDA has worked out a path for AI-powered medical devices that has its own elaborate documentation requirements. None of this is AI regulation in the EU sense. All of it is AI regulation in practice.

At the state level, things move faster. California, New York, Colorado, Texas, and Illinois have each passed or are advancing AI legislation that covers specific applications. California's law covers AI training data disclosure and automated decision-making. New York City's law on AI in hiring requires bias audits. Colorado's law applies to high-risk AI used in consequential decisions and largely mirrors the EU framework's logic. The picture is a patchwork that becomes painful for any company operating in multiple states, which is most companies.

The pragmatic answer for most US-based companies in 2026 is to use the EU AI Act framework as your internal standard. It is the most demanding, so compliance with it tends to cover the US patchwork by accident. Treat it as the floor of your discipline. Then have your lawyers add the state-specific extras.

What compliance actually requires, in practice

If you are in the high-risk tier, the regulatory framework asks you to do roughly seven things. None are individually complicated. Adding them all up produces real work.

Technical documentation

You must be able to describe, in writing and in detail, what your system does, what data it was trained on, how it was tested, what its known limitations are, how it has been validated, and what controls are in place. This is sometimes called a model card or a system card. It is not optional. It is the document the regulator will ask for first when the call comes.

Most companies discover, when they sit down to write the first model card for a system that is already in production, that they cannot. The training data is undocumented. The testing was informal. The controls are implicit in the engineer's head, not written down. The fix is to write the model card before you ship, not after the regulator asks. It also tends to surface problems you would otherwise have missed.

Risk assessment

For each high-risk system, you must perform and document a risk assessment. What could go wrong. What is the impact if it does. What are the safeguards. How will you detect the failure. This is similar to the safety engineering done in regulated industries for decades. It is new to most software companies and uncomfortable to do for the first time. Worth doing anyway.

Human oversight

There must be a human in the loop for consequential decisions, with the authority and the actual capability to override the AI. Both halves matter. A human who exists on paper but is given five seconds to review a decision the system has flagged is not real oversight. A human who has the time and the training and the authority to actually intervene is. Design for the second kind. Audit for the second kind. Do not let it drift toward the first.

Logging

Every decision the AI makes, you log. Inputs, outputs, confidence, who reviewed it, what the human decided in the end. This sounds like a lot of data, and it is, but it is the data you need both for compliance and for detecting your own failures. Treat the logging as the foundation of every other compliance activity.

Bias and fairness testing

For systems that affect people, particularly in hiring, credit, healthcare, and law enforcement, you must test for differential impact across protected groups, document the results, and demonstrate mitigation where you find problems. The tools for doing this have matured. The discipline of doing it consistently has not. Build the discipline.

Post-deployment monitoring

The system is monitored in production. Accuracy is tracked. Drift is detected. New failure modes are caught and fed back into the testing process. The model is not shipped and forgotten. It is shipped and watched.

Incident response

When the system fails in a way that causes harm, you have a documented process for responding. Notify affected users. Notify regulators where required. Investigate. Document what went wrong. Update the controls. This is the same incident response you have for security breaches. Extend it to AI.

The role nobody had three years ago

A new role has emerged at most companies serious about AI. The title varies. AI risk officer. Head of AI governance. Chief AI ethics officer. The role does not. It is the senior person whose job is to make sure your AI systems do not get you into trouble, and whose authority is to stop a launch when they will.

This role does not belong in engineering. The temptation, when companies first create the role, is to give it to the head of ML or the head of data science. This is a mistake. The same person cannot both build the AI and certify the AI as safe. The role belongs in legal, or in risk, or in compliance, or as a direct report to the CEO. It needs the authority to say no to engineering. If it does not have that authority, it is theatre.

The seniority matters. A junior compliance analyst with the AI risk officer title will not be able to override a senior engineer in the pre-launch meeting. The role needs to be senior enough that the conversation is between peers. At larger companies it may be a Vice President-level position. At smaller companies it might be the General Counsel adding AI risk to their portfolio. What matters is the authority, not the title.

Where the shadow AI problem lives

Every leader I talk to about AI governance eventually arrives at the same anxiety. They have inventoried the AI systems their engineering teams have built. They have a reasonable plan for those. Then they realise they have no idea what AI tools the rest of the company is using.

Marketing is putting customer data into consumer-grade AI assistants to draft emails. HR is using an AI screening tool a vendor sold them last quarter. Customer support is using an AI summarisation tool that came bundled with their helpdesk software. Sales is using an AI note-taker in every meeting. Finance is using an AI forecasting plugin in their spreadsheet.

None of this was a deliberate AI deployment. It happened the way all SaaS adoption happens. Someone signed up for a tool. It was useful. Word spread. By the time anyone with governance authority looked, half the company was using AI tools that had not been reviewed, that did not have data agreements in place, and that nobody could draw on the architecture diagram.

This is the shadow AI problem, and it is the part of governance most leaders are most behind on. The mitigation is not to ban all third-party AI tools, which would be impossible to enforce and would just push the use underground. The mitigation is to inventory honestly what is actually being used, to set clear policies on what data can go to which tools, and to give employees an approved set of tools that meet your standards. The approved set should be generous enough that employees do not feel the need to go around it.

> "The first audit of your shadow AI usage will surprise you. The second audit, three months later, will surprise you in a different direction. The shadow grows whether you watch it or not."

A composite case

A financial services company I advised in 2025 had been using AI in their credit decisions for two years. They were sophisticated, by the standards of two years ago. They had model cards. They had a fairness testing process. Their engineering team was proud of their work.

Then their general counsel attended an industry conference on AI regulation. She came back with a list of seventeen things she could not answer about how their AI complied with the new framework. The team had documentation, but the documentation was for engineers, not for regulators. They had fairness testing, but they could not produce a complete audit trail of testing decisions. They had post-deployment monitoring, but it was alerting their engineers, not their compliance team. Every individual piece existed. None of them was wired up for the kind of audit that was now plausible.

The company spent six months building what they called their AI governance program. Concretely:

- They hired an AI risk officer, reporting to the General Counsel, with authority to halt launches. The hire was a senior person, paid at parity with the senior engineering leads, deliberately.
- They wrote a regulator-facing version of every model card, separate from the engineering documentation. Same facts, different audience.
- They built a quarterly AI audit, where every deployed system was reviewed against the same checklist by someone independent of the team that built it.
- They inventoried the shadow AI, found seven third-party tools nobody had approved, evaluated them, banned two, approved five with conditions.
- They wrote an incident response playbook for AI failures, parallel to their existing security incident playbook.
- They updated their pre-launch checklist so no AI system could ship without an explicit sign-off from the AI risk officer.

None of this was glamorous. All of it was, in their general counsel's words, the price of being allowed to operate at all in two years.

Eighteen months later, they were one of three companies in their sector named in a regulator's list of best practices. New business came in partly because their compliance reputation made them an easier vendor for risk-averse customers. The cost of the program was real but recoverable. The cost of not having had it would have been worse.

Objections you will hear

"This will slow us down"

Yes. It will. The question is what alternative you are comparing against. Shipping fast and getting fined seven percent of revenue is not faster. Shipping fast and being unable to enter the EU market is not faster. Shipping fast and having a regulator force you to take down a feature is not faster. Compliance is overhead until you need it, and then the lack of it is catastrophic. Build the discipline early.

"We are too small for regulators to care about"

Sometimes true today, rarely true tomorrow. Regulators have been more interested in large companies first, but the laws apply to companies of all sizes. As tools mature and complaint volumes grow, smaller companies will get caught up. The cost of compliance is proportional, in most cases, to the complexity of your AI use, not to your company's revenue. Build it as you go and it stays manageable. Add it under regulatory pressure and it does not.

"Our lawyers will figure it out when the time comes"

Your lawyers cannot figure it out without your help. Compliance for AI requires technical documentation that only your engineers can produce, and risk assessments that only your product team can perform. The lawyer is a translator between your facts and the regulatory framework. If you do not have the facts, the lawyer has nothing to work with. Build the documentation muscle now, while the deadlines are still soft.

"We do not operate in the EU"

Yes you do, if any of your customers are EU citizens or any of your servers are in EU jurisdictions or any of your tools are accessed by EU users. The EU AI Act has extraterritorial reach for systems whose outputs are used in the EU. "We do not operate there" is, for most online businesses, an inaccurate description of the actual situation. Get the legal opinion before you bet the company on the assumption.

Questions for your leadership team

71. Do we have a current inventory of every AI system we have deployed, with its EU AI Act risk tier classified?
72. Who is our AI risk officer, what is their authority to halt a launch, and where do they report?
73. Do we have a regulator-facing model card for every high-risk system, separate from the engineering documentation?
74. When was our last audit of shadow AI usage across non-engineering teams? What did we find?
75. Do we have a documented incident response process for AI failures? When did we last rehearse it?
76. If a regulator wrote to us tomorrow asking for our compliance documentation, could we produce it within thirty days?

SIX THINGS TO TAKE FROM THIS CHAPTER

AI is no longer ungoverned. It is governed by law, with serious fines, and the laws have teeth.

Use the EU AI Act framework as your internal standard. It is the most demanding. Compliance with it covers most other jurisdictions by accident.

Build the documentation before you ship, not after the regulator asks. The exercise will improve the system regardless.

Create an AI risk officer role with real authority to halt launches. Place it outside engineering.

Inventory your shadow AI. It is bigger than you think. The mitigation is not a ban; it is an approved set of tools.

Compliance is overhead until you need it, and then the lack of it is catastrophic. Build the discipline before the deadlines turn hard.

ONE-LINE TAKEAWAY

The regulatory transition for AI has already happened. Most leaders are operating as though it has not. The work is unglamorous and the cost is real, but the alternative is a regulator's letter that you cannot answer. Build the governance now, while the rules are still being learned. It will be cheaper than building it under pressure later.

CHAPTER 13

Ethics and Responsibility Are Not the Same Thing

A CEO said something to me at a dinner last year that has stuck with me.

She told me, with some pride, that her company had "responsible AI." She listed the things they had built. A model review committee. A documented approval process. Required sign-offs from legal and security before any AI system went into production. A monthly audit. She had paid a consulting firm a substantial amount to design the whole thing.

I asked her one question. "Has the AI ever been wrong in a way that harmed a customer?"

She thought about it. "I am not sure. We have not really looked."

She had built one half of the picture. The half about accountability and process and who-signs-off. The half I would call responsible AI. She had not built the other half. The half that asks what the AI is actually doing to the people it touches. The half I would call ethical AI. And she did not yet know that these were two different things.

This chapter is the explanation I wish I had given her at that dinner. The point is small. The implications are not. Most leadership conversations about AI conflate ethics and responsibility, and the conflation is itself the source of much of the harm.

Two questions, not one

Here is the distinction in two sentences.

> "Responsible AI asks: who sees this, who decides this, who is accountable when this is wrong? Ethical AI asks: what does this system actually do to the people it affects, and is that right?"
>
> Sarah

Responsible AI is a question about who. Ethical AI is a question about what. They both matter. Neither one is the other. And a leader who builds only one of them has built half a program, even if the half they built was expensive and looks impressive on a slide.

This sounds like a small semantic distinction. It is not. It is the distinction that decides whether your AI program is genuinely safe or only performatively safe. So let me work through each one carefully, with examples of what each one looks like done well, and what each one looks like done in name only.

Responsible AI: who

Responsible AI is the procedural and governance side of the picture. The framework, the process, the people, the documentation, the trail of decisions that lets you reconstruct, after the fact, what was approved, by whom, and on what basis.

A company doing responsible AI seriously can answer questions like these without having to guess.

- Who reviewed the training data for this model before it shipped?
- Who has access to the logs of what the model has done in production?
- Who has the authority to take the model offline if something goes wrong?
- Who signed off on the deployment, on what date, with what conditions?
- Who is accountable, by name, if this system harms a customer?

Each of these is a who question. Responsible AI is the discipline of having a clear, documented, human answer to every one of them, before the system goes live and not after.

This is not easy work, but it is mostly mechanical work. It can be designed. It can be documented. It can be audited. It can be required as a pre-launch checklist. Companies that decide to do responsible AI well can usually get there in a year, given budget and senior attention. The previous chapter on governance was largely a chapter about responsible AI, though I did not call it that at the time.

What does responsible AI look like done poorly? Process theatre. A committee that meets monthly but cannot actually halt a launch. A documentation requirement that produces ten-page model cards no one ever reads. A sign-off that becomes a rubber stamp because the engineer presenting always wins the meeting. A risk officer who exists on the org chart but reports to the head of engineering, which means they will be overruled the first time it matters. These are the failure modes that look like responsible AI from the outside and operate as cover for engineering decisions on the inside.

Ethical AI: what

Ethical AI is the substantive question. Not who decided, but what was decided. Not what process was followed, but what the system actually does to the people who encounter it. And whether that is right.

A company doing ethical AI seriously asks questions like these, and is willing to live with uncomfortable answers.

- When this credit-scoring model rejects an applicant, what is the rejection rate broken out by age, race, gender, and zip code, and are the differences explainable by anything other than the protected attribute?
- When this hiring model ranks candidates, what features is it actually using, and would the

company stand behind those features if they were made public?

- When this medical model recommends a treatment, what kind of patient was it trained on, and does that match the kind of patient who is about to receive the recommendation?
- When this content moderation model removes a post, what voices is it systematically silencing, and is that the silencing the company actually believes in?
- When this chatbot tells a customer something wrong, what is the worst version of "wrong" we are willing to accept, and is our system designed to avoid the worst, or merely the obvious?

Each of these is a what question. Notice how different they are from the who questions. The who questions are about process. The what questions are about substance. The what questions cannot be answered by adding more sign-offs. They can only be answered by actually looking at the system's behaviour, measuring its outputs, and making a value judgement about whether those outputs are acceptable.

This is harder work. It cannot be designed once and then run as a checklist. It requires people who are willing to look at uncomfortable data, name the harm they see, and push back even when the system technically works. It requires a willingness to say no to launching a system that passes every governance check but does something the company would not be proud of.

Ethical AI is the half of the picture most companies underinvest in. The reason is partly that ethical AI is uncomfortable, because the answers are sometimes that you cannot ship the thing you have built. The reason is partly that ethical AI is unfashionable, because the metrics are softer than "sign-off completed." And the reason is partly that ethical AI requires people who can think clearly about values, which is a different skill from people who can think clearly about process, and the second skill is easier to hire for.

What goes wrong when you have only one

I have watched both failure modes. They are different, and both are real.

Responsible without ethical

This is the CEO at the dinner. The company that has built the process, the committee, the documentation, and the audit trail. Every sign-off is in place. Every model card is filed. The legal team is happy. The compliance team is happy. The board is reassured.

Then the AI goes ahead and quietly does something the company would be embarrassed by, if anyone had thought to look. A hiring tool that systematically rejects candidates from certain universities, traceable to a quirk of the training data nobody noticed. A medical triage tool that under-prioritises symptoms common in women because most of the historical training data was from male patients. A customer support chatbot that is rude to customers who type in non-standard English.

In each case the governance program did its job. Reviews happened. Sign-offs were given. Audits found nothing because the audits were not designed to find what was actually wrong. The harm was real but invisible to a process that was looking at the wrong things.

This failure mode is the more dangerous of the two, because the company believes it has done the work. The leadership team can answer every question a board might ask. The lawyers can answer every question a regulator might ask. And nobody asks the actual question that matters, which is, are we doing right by the people we are affecting.

Ethical without responsible

The opposite failure, less common but more familiar in academic and research settings. A team with strong ethical instincts, careful attention to fairness, deep concern for the people the system will affect. They have the right values. They are thinking about the right what questions.

What they do not have is the institutional machinery to operationalise their values. The decisions are made in conversations, not in documents. The reasoning is not written down. When the team changes, the values walk out the door with them. When a senior engineer pushes back on a fairness concern, there is no documented process for who decides, so the conversation devolves into seniority and personality. When a regulator asks for documentation, there is none.

This failure mode is less likely to cause large public harm, because the people involved generally have the instincts to avoid the worst outcomes. It is more likely to cause slow decay, as the ethical instincts get diluted over time by the operational pressures of a growing organisation, with no formal structure to hold the values in place.

Why this distinction is uncomfortable for leaders

Responsible AI is buildable. You can hire a consultant, design a governance framework, train your people, and check the box. It is real work, but it is the kind of work senior executives know how to commission. It produces deliverables. It produces a slide that says we are taking AI seriously.

Ethical AI is not buildable in the same way. It is a discipline that requires ongoing judgment, not a project that you complete. There is no consultant who can come in and install ethical AI. There is no slide that says we are done. The questions keep returning, in different forms, every time a new use case comes up.

This is why ethical AI gets short-changed in most companies. Leaders prefer projects with completion dates. Ethical AI does not have one. The leaders who do this well treat it less as a project and more as a permanent operational practice, like security or financial controls. The work is never done. The discipline is the deliverable.

How to actually do both

I am going to be more concrete here than is usual in writing about AI ethics, because the abstract version is what every leader has already heard and it has not changed behaviour.

For responsible AI, build the machinery

The previous chapter on governance covered most of this. Inventory your AI systems. Classify them by risk. Document them. Require human oversight for high-stakes decisions. Log everything. Audit regularly. Have a documented incident response process. Create an AI risk officer role with real authority. None of this is novel. Most of it is borrowed from other regulated industries that have done this for decades. Build it.

For ethical AI, build the practice

This requires a different set of habits, none of which substitute for the others.

First, measure the system's actual behaviour, not just its intended behaviour. For every consequential AI system, run the outputs broken down by demographic groups, by edge cases, by adversarial inputs. Do this before launch and on a recurring schedule after launch. Be willing to act on what you find.

Second, have an explicit values document, written by the leadership team, that says in plain language what you will and will not do. Not principles in the vague sense. Specific commitments. "We will not use AI to make final hiring decisions without a human in the loop." "We will not deploy facial recognition in our retail stores." "We will not use customer data to train models without explicit consent." The document is short, the language is specific, and the document is binding.

Third, give someone the authority and the courage to say no. Not the same person as the responsible-AI risk officer, necessarily, though sometimes the same person can hold both. The ethical AI lead is the person who can stop a launch on the grounds that, technically, this system is acceptable, but in practice, it is going to do harm we cannot defend. This is a senior, difficult role. The companies that fill it well are visibly more thoughtful than the companies that do not.

Fourth, listen to people outside your team. Internal teams develop blind spots about their own products. External advisors, civil society organisations, affected users, academic researchers, can see what you cannot. Build the practice of seeking outside perspective before the system ships, not after a critic forces the conversation.

Fifth, be willing to ship less, slower, smaller. The ethical AI program that never says no is not an ethical AI program. It is a permission system that happens to be called ethics. If your team has never killed a project on ethical grounds, the discipline is not real.

The hard cases, briefly

Most ethical AI conversations get derailed by the easy cases. Of course we should not have AI systems that discriminate. Of course we should not deploy AI that violates privacy. The hard cases are the ones where reasonable people disagree, and the work of ethical AI is to develop the company's position on the hard cases before they become crises.

A few examples of the hard cases, deliberately not resolved here, because the resolution depends on the company's specific context and values.

- If your model improves average outcomes for everyone but improves outcomes more for some groups than others, is the differential improvement a problem you must fix?
- If your AI catches fraud better than humans but generates false positives that disproportionately affect certain communities, what error rate is the right trade-off?
- If your customers explicitly ask for an AI feature you believe is harmful to them, do you build it because they asked, or refuse because you believe you know better?
- If your competitor ships a feature you believe is unethical, do you ship the same to stay competitive, or refuse and risk losing market share?
- If using your customers' data to train your AI would dramatically improve the product for those same customers, but they did not anticipate this use when they signed up, is the implicit consent enough?

Every one of these is a real decision real leadership teams have faced. None of them has a clean answer. The work of ethical AI is to think about them carefully, develop the company's position, write the position down, and live by it consistently. Companies that have done this work are visibly different from companies that have not, even when both have responsible AI governance in place.

A composite case

An insurance company I advised in 2024 had built what they believed was a state-of-the-art responsible AI program. They had a committee. They had documentation. They had a risk officer. Every claim-handling AI was reviewed quarterly. They were genuinely proud of the program, and the General Counsel had presented it at industry conferences.

In late 2024 a researcher contacted them with a finding. Their fraud-detection AI was flagging claims from neighbourhoods that were predominantly low-income for human review at three times the rate of claims from wealthier neighbourhoods, even controlling for claim size and policy type. The pattern was visible in the public data the researcher had access to. The company's internal audits had not caught it, because the audits were checking whether the right people had signed off on the model, not whether the model was producing fair outcomes.

The conversation that followed inside the company was harder than the technical fix. The fraud-detection model was, in a narrow sense, working. Most of the flagged claims did turn out to be questionable. The model was improving the company's loss ratio. The team that had built it pointed out, accurately, that they had followed every step of the responsible AI process.

The General Counsel, to her credit, pushed back hard. She said, in a meeting I was in, "We did what we said we would do. We are still doing harm. Those are both true. We have to fix both halves."

The company spent the next six months doing the second-half work. They added fairness testing to every consequential model, not as a one-time audit but as a recurring metric tracked on the same dashboards as accuracy. They added an external review board with two academic ethicists and a civil society representative. They wrote a values document that committed them to specific things, including ones that constrained their commercial freedom. They retrained the fraud model with explicit constraints on demographic differential impact, accepting a small reduction in raw fraud detection rate in exchange for a more equitable distribution of human review.

Eighteen months later, the new program had caught two more fairness issues in other models. Both were fixed before any external researcher noticed. The company's General Counsel told me that the most important learning was that the original program had been responsible without being ethical, and that the gap had taken a researcher's letter to make visible. "We thought we were doing the work," she said. "We were doing half of it. And the half we were doing was the half that was easier to do."

Objections you will hear

"We do ethical AI. We have a code of conduct."

A code of conduct is a useful document. It is not ethical AI. Ethical AI is measured by what your systems actually do to people, not by what you have written about what they should do. The code is a starting point. The behaviour of the systems is the measurement. If you have the first and not the second, you have aspirations, not practice.

"This will paralyse our team. They will be afraid to ship anything."

If your team becomes paralysed because someone is now asking ethical questions, the problem was not the questions. The problem was that the team was previously shipping without anyone asking. The mature version of ethical AI does not paralyse teams. It makes them think more carefully about a smaller set of systems and ship those with confidence. Teams that have done this work for a year are usually faster than they were before, not slower, because they avoid the rework that comes from launching something problematic.

"We are too small for ethical AI as a discipline"

You are not. The discipline scales down. A small company does not need a five-person ethics team. It needs one senior person who asks the what questions in every product review. The smaller the company, the more important that question is, because there are fewer other checks on what gets shipped.

"Our competitors will eat our lunch while we are arguing about ethics"

Sometimes true in the short term. Almost never true in the long term. The companies that ship ethically problematic AI usually have to walk it back later, and the walking-back is more expensive than the original would have been if done thoughtfully. The reputation cost of being the company in the news story is larger than the speed cost of pausing to think. Slow is sometimes fast.

Questions for your leadership team

77. Can we describe, separately and in our own words, what we mean by responsible AI and what we mean by ethical AI? Do those definitions sit in the same document?
78. For each of our high-stakes AI systems, when was the last time we measured the outcomes by demographic group? What did we find? What did we do about it?
79. Do we have a written values document that says specifically what we will and will not do with AI? Is it binding when it inconveniences a launch?
80. Who in our organisation has the authority to halt a launch on ethical grounds, and have they ever used it? If never, are we sure no launch should have been halted?
81. How often do we listen to perspectives from outside the team, before a system ships, not after a critic forces the conversation?
82. If a researcher contacted us tomorrow with evidence that one of our AI systems was harming a group of customers, would we know how to investigate, who is accountable, and what we would do?

SIX THINGS TO TAKE FROM THIS CHAPTER

Responsible AI is about who. Ethical AI is about what. They are not the same. A program that builds only one has built half.

Responsible AI can be designed once and operated as a process. Ethical AI is a permanent practice, not a project with a completion date.

Responsible-without-ethical is the more common and more dangerous failure mode. The company believes it has done the work; the harm is invisible to processes looking the wrong way.

Measure outcomes by demographic group. Track the measurements. Be willing to act on what you find.

Write a specific values document that constrains your commercial freedom in named ways. Vague principles are not ethics.

Give someone the authority and the courage to halt a launch on ethical grounds. If no launch has ever been halted, the discipline is not real.

ONE-LINE TAKEAWAY

The leader who builds responsible AI without ethical AI has built a process that produces compliant harm. The leader who builds ethical AI without responsible AI has built good intentions that cannot survive their author leaving. Build both, distinctly, and treat each as the discipline it is.

CHAPTER 14

The Bill That Surprised the CFO

A CFO put a printout on the table at a board meeting last summer.

It was the company's AI bill for the quarter. The number was bigger than the engineering team's salaries for the same period. He looked at me. He did not look angry. He looked confused.

"I do not know how to think about this," he said. "Six months ago this line item was zero. Now it is the third-largest line on this page. Nobody in finance can tell me whether that is good or bad."

The thing is, neither could the AI team. They could tell him the cost per query. They could not tell him whether the cost per query, multiplied by the volume the company was now serving, was a reasonable price for the value being delivered. The financial language for what they were buying did not yet exist at most companies. It still does not, fully. This chapter is the language.

Why the bill keeps surprising people

AI economics breaks the financial intuitions most leaders have built up over decades. Three things break at once.

The first is that the unit cost is small but the volume is huge. A query that costs a tenth of a cent looks like a rounding error in the budget meeting. The same query, run a million times in a successful month, is a thousand dollars. Run a hundred million times in a viral month is a hundred thousand dollars. The cost scales with usage in ways most product budgets are not built to handle. Software companies are used to fixed-cost development and near-zero marginal cost of serving. AI inverts that. The marginal cost of serving is real and meaningful.

The second is that the cost curve is moving fast. The same query that cost a penny in 2023 costs a tenth of a cent in 2026. By 2028 it may cost a hundredth of a cent. A financial analysis done in January is stale by July. Most CFO models assume cost is roughly stable over a planning horizon. AI cost is dropping by a factor of ten every twelve to eighteen months and shows no sign of stopping. Plan accordingly.

The third is that the costs are split across categories that finance departments do not yet have lines for. There is the inference bill from the model provider. There is the infrastructure cost of running your supporting systems. There is the cost of the engineers who maintain it. There is the cost of the data labelling team you suddenly need. There is the cost of the compliance program from the governance chapter. Add all these up and the question "how much does our AI cost?" becomes harder to answer than it looks.

Where the money actually goes

Let me break down the cost categories the way I would explain them to a CFO who has just slid a printout across the table.

Inference

The cost of running your model in production. Pay-per-query when you use a hosted model from a vendor. Pay-for-GPU-time when you run your own. This is the line item that tends to grow most dramatically as the product succeeds, and it is the line item most often underestimated in the original business case.

A reasonable mental model is that inference cost for a frontier model in 2026 is something like a thousand dollars per million tokens of output, give or take an order of magnitude depending on which model. A typical query consumes a few thousand tokens. So a frontier-model query costs a few cents. A small-model query is closer to a few tenths of a cent. Multiply by your volume to get the line on the bill.

Training and fine-tuning

If you fine-tune your own models, the training runs are expensive. Tens of thousands of dollars per run is typical for a meaningful fine-tune, and you will do many runs as you iterate. If you train models from scratch, the costs jump to millions per run, and very few companies should be doing this. If you use only the off-the-shelf models, this line is zero, which is one of the under-appreciated reasons to default to prompting and RAG rather than fine-tuning.

Data labelling and curation

Often the largest hidden cost. To fine-tune a model you need labelled examples. To run an evaluation harness you need carefully written test cases. To do supervised learning you need ongoing labelling as your data drifts. The labelling can be done in-house, which is slow and expensive, or outsourced, which is faster and still expensive. A serious AI project at scale spends substantial money here. Most teams do not budget for it at the start and then discover, six months in, that their progress is bottlenecked on labelling rather than on engineering.

Infrastructure

The cost of all the supporting systems from the stack chapter. The vector database. The orchestration. The prompt management. The evaluation harness. The observability. None of these are catastrophically expensive on their own. All of them together are a meaningful line in the budget, and most of them are billed monthly, which means the cost recurs whether or not anyone is using the system.

People

Over a three-year horizon, by far the largest cost. The AI engineer. The ML engineer. The data engineer who keeps the pipelines clean. The AI risk officer from the governance chapter. The compliance lead. The data labelling manager. The customer support engineer who handles the AI-specific tickets. Easy to add half a dozen specialised roles, each of whom is expensive in the 2026 market.

This is the line that catches CFOs by surprise most often. They look at the inference bill and worry about that. The inference bill is real, but it is rarely the largest cost in the long run. People are.

Compliance and risk

New as a line item, but real and growing. The cost of the governance program. The cost of legal review. The cost of the documentation and the audits and the model cards. The cost of the bias testing. The cost of the third-party audits some regulators will require. None of this is enormous individually. Together it is a meaningful percentage of the total program cost, and unlike the others, it does not stop growing as the program matures.

The cost curve, and what to do about it

Inference cost has dropped by roughly tenfold per year for the past three years. The trend looks likely to continue for several more, driven by better hardware, better algorithms, more competition between model providers, and increased efficiency of inference engines.

This means two things for your financial planning.

First, do not let the current cost of a use case decide whether you pursue it. The thing that is too expensive today will be affordable in a year. The CFO at the meeting I described had killed a project six months earlier because the inference cost looked prohibitive. By the time he was looking at the bill that surprised him, the cost per query had dropped enough that the killed project would have been comfortably profitable. He had made the right decision with the information he had. He had also made the wrong decision in hindsight, because the rule of thumb of "costs are roughly stable" had failed him.

Second, expect your competitors to do this calculation as the cost falls. Use cases that look uneconomic to you today are looking the same way to them. As the cost drops past their threshold, they will deploy. So will you. The question is who gets the right product ready to deploy first, not who has the cheapest cost-per-query in a given quarter.

Practical advice: build the use case quietly while the costs are still high. Have it ready to launch when the math turns. The companies that launch at the moment the math turns capture the market. The companies that wait until they see competitors launch are six months behind.

The pricing model trap

Every CFO I work with eventually asks the same question. How do we charge customers for AI features without losing money?

The honest answer is that nobody has fully figured this out. The dominant pricing models in 2026 fall into three camps, and each has a failure mode.

Per-user pricing

Charge a flat monthly fee per user, regardless of how much they use the AI. Easy to explain. Easy to forecast. Vulnerable to the heavy-user problem, where a small percentage of users consume the majority of the AI capacity and become unprofitable, while light users effectively subsidise them. Almost every per-user AI feature I have looked at carefully has a heavy-tail usage distribution that the original pricing did not anticipate.

Per-query pricing

Charge for each AI call. Lines up cleanly with your underlying inference cost. Hard for customers to predict their bill, which they hate. Tends to cause customers to under-use the feature, which reduces the value they get from it, which reduces their willingness to renew.

Hybrid pricing

Base fee plus usage above a threshold. The model most companies eventually arrive at. Combines the predictability of flat pricing with the protection of metered pricing. Operationally complex. Harder to explain. Often the right answer.

The trap most companies fall into is committing to per-user pricing in year one because it is easy, then discovering in year two that their unit economics are upside down. The migration to a hybrid model is painful because existing customers will resist the change. Plan for hybrid from the start if your AI feature has meaningful per-query costs.

ROI, and the lies people tell about it

Most AI ROI claims in 2026 are not honest. They are not lies, exactly, but they are not the carefully-measured business cases a CFO can take to a board. They are estimates, often optimistic, often without a counterfactual, often without the costs from the categories above included properly.

Here is how to do this honestly.

First, name the specific benefit in measurable terms. Not "productivity improvement." Specifically, "reduces time to resolve a customer support ticket from 18 minutes to 9 minutes, on the 40 percent of tickets the AI can handle, at our current volume of 50,000 tickets per month." The specificity is the discipline.

Second, measure the actual baseline before the AI was deployed. Not a remembered baseline. A measured one. Companies routinely overestimate how much time something took before they automated it, partly because their memory of pre-automation life is the painful version. Real measurement, before the change, is the only honest comparison.

Third, include all the costs from the categories above. The inference cost is the smallest of them. The people cost is the largest. If your ROI calculation only includes the inference cost, it is wrong by a factor that matters.

Fourth, run the math at three volumes. The current volume. Twice the current volume. Ten times the current volume. The economics that work at one volume can fail catastrophically at another. The CFO needs to see how the math changes as the product succeeds.

Fifth, include the counterfactual. What would have happened without the AI? Sometimes the team would have hired more humans to do the work, which has costs. Sometimes the team would have left the work undone, which has different costs. The counterfactual matters because the AI did not save you the full cost of doing the work; it saved you the difference between doing it with humans and doing it with the AI.

Most AI ROI cases I see fail at one of these five steps. Doing all five honestly produces smaller, more defensible numbers, which are also numbers a CFO can actually plan against.

> "If your AI ROI case looks like it returns 1000 percent in year one, the case is wrong. Real returns on real AI projects are in the 20 to 80 percent range. Anything bigger is usually a measurement error."

When AI is not economic, and probably never will be

Some use cases never close on the math, no matter how far the cost curve falls.

Use cases where the cost of being wrong is enormous and the cost of being right is low, like AI making final decisions on consequential things with no recourse, will never have economic AI that is also safe AI. The unit economics may work; the risk economics will not.

Use cases where the volume is too small to amortise the setup cost, like a niche internal tool used by twenty employees a week, will rarely justify the engineering and operational overhead, even if the inference is free.

Use cases where a deterministic solution already exists and works, like calculating sums, looking up exact records, applying clear business rules, do not need AI. The fact that you can use AI does not mean it is cheaper than a database query that has worked for forty years.

Use cases where the human time being replaced is already cheap, like volunteer moderation or unpaid user contributions, will not pencil out against the cost of running the AI.

The leader who insists on AI for every problem because AI is the fashionable answer of the decade is the same leader who insisted on blockchain in 2018 and cloud in 2012. Some problems are AI problems. Most are not. The discipline of saying no to use cases where the math does not work is part of the job.

The capacity problem

One more cost dynamic worth understanding, because most companies do not.

If you rent inference from a hosted model provider, you pay per query. The costs scale linearly with usage. This is the most common model and the easiest to budget.

If you commit to a reserved capacity contract for better pricing, you are now paying for capacity whether you use it or not. The pricing per query is lower, sometimes substantially. The risk is that you over-commit. Reserved capacity that goes unused is pure waste.

If you run your own models on your own GPUs, you have made a fixed-cost commitment. The hardware is paid for upfront and depreciates over years. The cost per query falls to nearly zero at sufficient volume. The risk is again over-commitment, and additionally, GPUs depreciate quickly as better hardware appears.

The right answer depends on your volume and its predictability. Low and variable: pay-per-query. High and steady: reserved capacity. Very high and growing: consider owning the GPUs. Most companies should be in the first or second bucket. The third is a serious commitment that few use cases justify.

A composite case

A media company I worked with in 2025 built an AI feature to summarise news articles for their readers. The first business case showed the feature would cost an estimated three cents per summary at frontier-model prices, with an estimated daily volume of fifty thousand summaries, working out to roughly forty-five thousand dollars a month in inference cost. The product team was nervous about the number. The CFO approved the launch on the condition that the team would track actual costs closely.

Three things happened in the first six months that nobody had predicted.

First, the volume was much higher than expected, almost three hundred thousand summaries a day at the peak.

Users loved the feature. The inference cost climbed to over a quarter of a million dollars a month. Not catastrophic, but five times the budget.

Second, the cost per summary dropped during the same period. By the end of six months it was closer to half a cent. The model the team had originally been using had been replaced by a smaller, more efficient model that produced summaries the team and a careful evaluation could not distinguish from the original. The unit cost was now a sixth of the original estimate.

Third, the team had not budgeted for the supporting costs. They had hired an AI engineer to maintain the system, a data engineer to handle the article ingestion, and added prompt management and observability tools. The supporting costs were roughly forty thousand dollars a month, more than the original inference budget had been.

Net of all this, the feature was costing about a hundred and fifty thousand dollars a month against a budget of forty-five. The volume had grown six times. The unit cost had dropped six times. The supporting costs were almost as big as the inference. None of this had been visible in the original business case.

The CFO told me later that the financial education the project provided was the most valuable thing it produced. They had thought they understood AI economics. The first project taught them that they did not. The second project, six months later, had a much better business case, with the right cost categories, the right volume scenarios, and the right contingencies. It launched on time and within budget.

Objections you will hear

"We will save money replacing people with AI"

Sometimes. Often not. The people you would replace cost less in total than you think, once you include the supporting AI costs honestly. The people you would not replace, who continue to do the work the AI cannot do, become more important and harder to keep. The pure cost replacement calculation rarely lands as cleanly as the vendor demo suggests. Be skeptical.

"Our vendor said inference costs are negligible"

Inference cost is small per query. It is not small per year at scale. And it is one of five cost categories, not the only one. Vendors quote the cost they can quote. The total cost is yours to figure out, and it is bigger than the vendor's quote.

"We will lock in pricing with a long-term contract"

This sounds prudent. It is risky. Inference prices are dropping faster than any other software category. A two-year contract at today's prices may look smart in month three and ruinous in month eighteen. Negotiate shorter terms with options to renew at then-current pricing. Vendors will resist. Insist anyway.

"AI is too expensive for our use case"

Today, maybe. In nine months, probably not. The cost curve is steep enough that today's no can be next year's yes. Build the case quietly. Be ready to launch when the math turns. The companies that wait to see the math turn before they start building will be late.

Questions for your leadership team

83. Can we name, for each major AI program, the five cost categories (inference, training, data, infrastructure, people, compliance) and the actual numbers for each?
84. How sensitive is our unit economics to the success of the product? If volume grows ten times, do we make more money or lose more money?
85. What is our actual ROI calculation, including all costs and a measured baseline? Or are we using vendor estimates?
86. If our model provider raised prices by 50 percent tomorrow, what would happen to our unit economics?
87. Are we charging customers in a way that aligns with our underlying costs, or are heavy users subsidised by light users?
88. Which AI use cases have we said no to because the math did not work? When did we last revisit them?

SIX THINGS TO TAKE FROM THIS CHAPTER

AI cost has five categories. Inference is the smallest. People are usually the largest. Budget all five honestly.

The cost curve is dropping by roughly tenfold per year and shows no sign of stopping. Today's uneconomic use case is next year's product opportunity.

Per-user pricing for AI features has a heavy-tail problem.

Plan for hybrid pricing from the start.

ROI claims above 200 percent are usually measurement errors. Real AI returns are in the 20 to 80 percent range. The honest cases are the ones a CFO can plan against.

Some use cases will never be economic, no matter how cheap the inference gets. Knowing which is part of the discipline.

Vendors quote the cost they can quote. The total cost is yours to figure out, and it is bigger than what shows up on the vendor's invoice.

ONE-LINE TAKEAWAY

AI economics breaks the financial intuitions most leaders built up over decades. Costs that scale with usage, cost curves that drop tenfold per year, cost categories that finance has no lines for. The leaders who get this right do not have better forecasts. They have honest categories, measured baselines, and the discipline to revisit the math every quarter, because the math keeps moving.

CHAPTER 15

Staying Useful

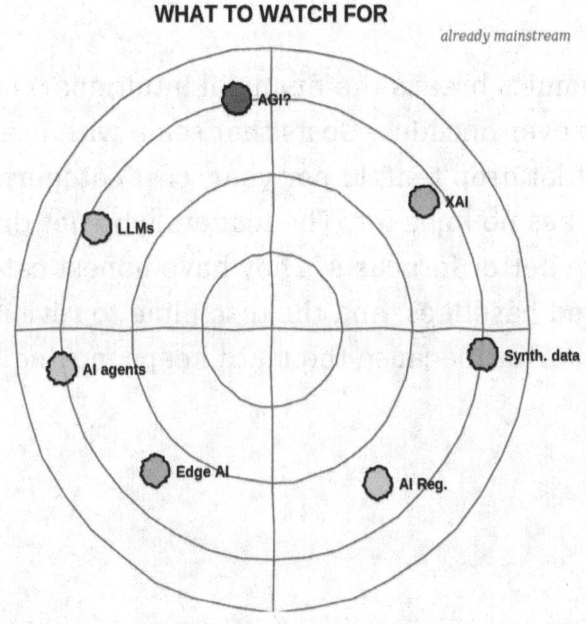

Every leader who has worked in technology for more than a decade has watched a once-dominant skill set become irrelevant. The COBOL programmer, the network engineer, the SAP consultant, the iOS developer, the data scientist who refused to learn deep learning. The shelf life of specialist skill is shorter than it has ever been, and shortening.

This chapter is about what to actually do with that fact. Not abstract advice about "continuous learning." Practical patterns for staying useful in a field that does not slow down.

Four trends to actually pay attention to

Many AI trends will turn out to be noise. A few will matter. Here are four that are worth your time.

Large language models become infrastructure

By 2026, large language models are no longer the exciting new toy. They are quietly becoming infrastructure, the same way databases or web frameworks did before them. The interesting work is not building new models from scratch. It is building applications on top of them. Most enterprises should be thinking about LLMs the way they think about cloud computing: as a utility to integrate, not a technology to invent.

Explainable AI moves from research to requirement

Regulators in finance, healthcare, hiring, and insurance are increasingly insisting that high-stakes AI decisions be explainable. "The model said no" is no longer an acceptable answer. Techniques like SHAP and LIME, attention visualisations, and inherently interpretable models are moving from research curiosity to operational requirement. Plan for this in any system that affects individuals.

AI agents become useful, slowly

The dream of an autonomous AI assistant that can complete multi-step tasks is older than most readers of this book. In 2026, the technology is finally usable for narrow, well-defined workflows. It is not yet usable for the kind of broad autonomy the marketing implies. Pilot agent-based workflows in low-stakes parts of the business. Do not bet the company on them yet.

AI regulation becomes real

The European Union's AI Act is in force. Similar frameworks are advancing in the United States and most major markets. The regulatory landscape for AI in 2030 will look more like the regulatory landscape for finance or pharmaceuticals than like the unregulated era we just left. Build your AI practice as if the regulation already existed. The cost of retrofitting compliance is much higher than the cost of building it in from the start.

How leaders actually stay sharp

A short list of habits I have stolen from people who have stayed at the front of technology for thirty years.

Spend two hours a week reading

Not Twitter. Not LinkedIn posts. Actual long-form writing: papers, essays, the better newsletters. Two hours a week, protected on the calendar. Over a year, this is a hundred hours of slow learning. The compounding effect over a decade is the difference between staying current and being left behind.

Talk to practitioners, not vendors

Vendors will tell you what to buy. Practitioners will tell you what works. Keep a list of three to five people in the field whose opinions you trust, and have a conversation with each of them once a quarter. Buy them dinner. The information asymmetry that exists between people who actually build systems and people who only sell or describe them is enormous, and the only way to close it is to talk to the builders.

Build something small yourself, occasionally

You don't have to become an engineer. But spending a weekend trying to build a small thing with the technology you are making decisions about will teach you more about its real strengths and limitations than any number of analyst reports. The CEO who has personally tried to prompt-engineer a useful workflow makes better decisions about generative AI than the one who hasn't.

Watch what teenagers do

This is not a joke. Technology adoption patterns in the next decade are most visible in how teenagers use technology today. If a behaviour is mainstream in a fifteen-year-old's life, it will be mainstream in your customer's life in five years. Pay attention.

Staying useful as an organisation

The same principles apply at scale.

Invest in learning, structurally

Most organisations claim to value learning and structurally penalise it. Time spent learning is time not spent shipping, and shipping is what gets measured. The fix is to make learning structural, dedicated time for it, dedicated budget, dedicated career rewards for people who teach others. Companies that get this right outlearn their competitors over a decade. The compounding is brutal.

Stay close to research, but not too close

Most enterprises should be one to two years behind the cutting edge of research, not on it. The cutting edge is unstable, expensive, and often the wrong technology by the time it stabilises. The sweet spot is to track research, pilot the things that look stable, and adopt them once they are proven but before they are commoditised.

Be honest about what you can absorb

Most digital transformation strategies fail because they exceed the organisation's actual capacity to absorb change. People can only learn so much, change so many habits, adopt so many new tools at once. The leader's job is to pace the change. Three priorities ruthlessly executed beat fifteen priorities loosely attempted.

> "The leaders who survive long technology transitions are not the ones with the best forecasts. They are the ones with the best learning habits."

A composite case

Two senior technologists I have known for fifteen years started in similar positions in similar companies in 2010. Both were strong engineers at the time. Both became CTOs within a few years.

One of them, the one I will call A, built a habit early in her career of reserving Friday afternoons for reading. Two hours, every Friday, for fifteen years. No meetings allowed. No email. Just long-form material: papers, books, the better newsletters. Over fifteen years that is roughly 1,500 hours of slow, deliberate learning. The compounding effect, I have watched, is substantial. She speaks with current vocabulary about every technology category. She catches vendor misrepresentation in real time. She mentors junior engineers in ways that draw on patterns spanning decades. Her companies tend to make good early bets and avoid bad ones. She is in demand.

The other, B, was always too busy. The Friday afternoon was never possible. There was always a meeting, a fire, a deadline. He read in airports and during commutes, which over fifteen years adds up to a different and shallower kind of learning. He is still a competent engineer. He has spent the last five years quietly falling behind in the categories where the field has moved fastest. He misses things in vendor pitches because he does not yet have the vocabulary. He has been passed over for two roles he wanted.

The difference between them is not talent. They started with comparable talent. It is a single calendar discipline maintained for fifteen years. Two hours a week. Eighty hours a year. Compounded across a career, the difference is the difference between being current and being left behind.

I told B this story once. He said, with some pain, that he agreed completely. Then he said he was still too busy. He may always be.

Objections you will hear

"I do not have time to read for two hours a week"

You probably have two hours a week you spend on something less valuable. Most leaders, when they audit their actual calendar, find at least four hours of meetings that could have been emails. Reclaim two of those for the reading block. The work is not adding two hours; it is choosing which two of the existing two hours to use this way.

"Industry conferences are the same content, repackaged"

Often true. The value of conferences in 2026 is rarely the talks. It is the conversations in the hallways with people doing the work. If you go to conferences for the talks, you are right that they are not worth it. If you go to talk to practitioners, with specific questions you have prepared in advance, the conferences can be among your highest-leverage learning of the year.

"Reading is a luxury, not a job requirement"

It is a job requirement that most companies fail to make explicit. Senior technical leaders who do not read steadily fall behind, and the falling behind is invisible until the year it catches up. The leaders who treat learning as core to the job, not as an extra, are usually the ones who are still leading effectively a decade in.

"Talking to practitioners is networking, which I hate"

Networking with strangers at events is one form. Talking with three or four practitioners you respect, over coffee, with a specific question, is another. The first is uncomfortable for most people. The second is among the most enjoyable parts of senior technical work. Distinguish them.

"I will catch up when things slow down"

They will not slow down. They have not slowed down for any senior leader I have known in twenty years. The discipline of learning under conditions of permanent busyness is the only discipline that actually exists. Build it now or accept that you will not build it.

Questions for your leadership team

89. When did each of us last read a substantial technical document we did not have to read for an immediate decision?
90. Who in our network would tell us hard truths about our AI work? When did we last actually call them?
91. If we left our current company tomorrow, would we still be a credible AI leader two years from now? What would have to be true?

92. What is one thing we believed about AI in 2023 that we now know was wrong? How did we learn? Could we learn the next wrong belief faster?
93. Are we spending our learning time on the right things, or on what is loudest? When did we last audit our reading diet?

IF YOU REMEMBER NOTHING ELSE FROM THIS CHAPTER

Two hours a week reading long-form, protected on the calendar. Over a decade, this is the difference.

Talk to practitioners more than to vendors. The information asymmetry is enormous.

Pace organisational change to what the people in the organisation can actually absorb.

Stay one or two years behind the research cutting edge, not on it.

ONE-LINE TAKEAWAY

Staying useful in a fast-moving field is a structural choice, not a heroic one. Build the learning habits, talk to builders, pace organisational change, and pay attention to the next generation's defaults.

CHAPTER 16

Your First Thirty, Sixty, and Ninety Days

On the Monday morning of your first week as an AI leader, you will feel two pressures at once.

The CEO will want to see momentum. You will want to demonstrate that hiring you was the right call. Both pressures will push you toward visible action. The temptation to announce something, kill something, build something, hire someone, anything that will look like progress in your first month, is enormous.

Resist it.

The first ninety days of an AI leadership role are the most consequential ninety days of the entire job. What you do in that window will shape the next two years, in ways that will be hard to unwind later. And the highest-leverage thing you can do during that window is, almost certainly, not what your instinct will tell you to do.

This is the closing chapter, and the most practical one in the book. It is the playbook I would hand to anyone stepping into AI leadership for the first time, structured as three thirty-day blocks. Read it through once now. Then read it again on your actual first Monday. Then again at day thirty and sixty and ninety, as a check on what you have done.

The frame: three different jobs, in sequence

The reason the ninety-day window matters is that you are doing three different jobs in sequence, and the order is not interchangeable.

In the first thirty days your job is to listen and map. Not to act. To understand the terrain you have inherited. To meet the people who will make or break your tenure. To find out what is actually happening with AI inside the organisation, which is almost never what the org chart says is happening. The goal of this month is not progress. It is a true picture.

In the second thirty days your job is to focus and govern. To take the picture you have built and use it to make the hard decisions about what stays, what changes, and what stops. To establish the governance machinery that will let you say no to bad projects and yes to good ones with the same authority. The goal of this month is a defensible set of priorities and the institutional weight to defend them.

In the third thirty days your job is to ship and learn. To launch the first deliberate work under the new system. To establish the rituals and reviews that will outlast your attention. To produce visible progress that you can stand behind, and that the rest of the organisation can recognise. The goal of this month is the first real wins, and a system that will keep producing them after you stop personally babysitting it.

Each of these jobs builds on the previous one. Skip the listening and your focus will be wrong. Skip the focus and your shipping will be premature. Do the three in sequence and the ninety days produce a foundation. Try to do them in parallel and the ninety days produce noise.

> "Most failed AI leadership tenures fail in the first thirty days, because the leader started acting before they knew what was true. The action looked decisive. It was just wrong."

Days 1 to 30: Listen and map

The temptation in month one is to announce a strategy. Do not. You do not know enough yet. The strategy you announce in week two will be the strategy you have to walk back in month four. Buy yourself the time to actually understand the picture.

Concretely, in the first thirty days, here is the work.

Inventory the AI systems

Build a complete list of every AI system the organisation has deployed, is building, or is using through a third party. Include the shadow AI from the governance chapter. Include the AI features bundled into SaaS tools nobody has reviewed. Include the experiments and prototypes that may or may not have gone to production.

Most companies, when they do this exercise for the first time, find roughly twice as many AI systems as their official records show. Do not be alarmed. Be precise. For each system, record what it does, what data it uses, who owns it, what risk tier it falls into, and what is known about its performance.

This document is the foundation for everything else you will do. It will be wrong on the day you finish it, because new systems will have appeared while you were writing. Update it monthly. It is the single most useful document an AI leader maintains.

Meet thirty people

In your first thirty days, have a one-on-one conversation of at least thirty minutes with at least thirty people. The list should include every engineer working on AI, every product manager whose product touches AI, your CFO, your General Counsel, your CISO, your Head of HR, your Head of Customer Support, and a sample of the customers who actually use the AI features in production.

In each conversation, ask three questions. What is working well that you want me to protect? What is broken that you want me to fix? What is being avoided because no one has the authority to act? Listen more than you talk. Take notes. Do not promise anything yet. The point is information, not action.

At the end of these conversations you will see patterns. The same complaint will surface in multiple meetings. The same project will be named as broken by three different people. The same opportunity will be flagged by people who have never spoken to each other. These patterns are your priority list. They are also, almost always, different from the priority list you would have written from your office on day one.

Read the documents

Find and read every document that already exists about AI in your organisation. The strategy decks. The board memos. The risk assessments. The model cards, if any exist. The compliance documentation. The vendor contracts. The customer complaints log filtered for AI-related issues.

Most of these documents will be incomplete, out of date, or contradictory. That is useful information. The pattern of what is documented well and what is documented badly tells you where the organisation is mature and where it has been operating on hope.

Sit with the data

If you can, spend at least a full day reading the actual logs of what your AI systems have done in production. Customer support transcripts the AI handled. Recommendations the AI made. Decisions the AI flagged. Errors the AI produced. This is uncomfortable work and most leaders skip it. The ones who do not skip it leave the day with a clearer picture of what their AI is than any briefing could have given them.

What not to do in month one

Do not announce a vision. Do not restructure the team. Do not fire anyone. Do not commit to a strategy. Do not promise the board anything specific. Do not sign any vendor contracts. Do not deploy any new AI systems. Do not pause any existing AI systems unless they are actively causing harm.

This will feel slow. It will feel like you are not doing your job. The CEO will start to wonder if you are too cautious. Tell the CEO, plainly, that month one is for listening and month two is for action. Buy yourself the time. The leaders who succeed in this role are the ones who use month one for what it is for.

Days 31 to 60: Focus and govern

By day thirty you should know things you did not know on day one. The inventory should be substantially complete. The patterns from your thirty conversations should be clear. The data you have read should have given you a feel for the gap between how the systems are described and how they actually behave.

Now you act, but selectively. Three workstreams in parallel.

Choose the focus, ruthlessly

From the inventory and the conversations, pick three to five things that matter most for the next year. Not ten. Three to five. These are the AI initiatives the organisation will invest in, the rest will be wound down, paused, or maintained at minimum cost.

The criteria for what makes the focus list: clear business value, alignment with strategy the rest of the company is pursuing, technical feasibility given current capabilities, acceptable risk profile, and someone competent who can own delivery. If a project fails any of these, it does not make the list. The projects that fail are not personal. They are just not the focus.

Write the focus list down. Share it with the CEO. Share it with the board. Share it with the affected teams. The act of writing it down is half the discipline, because it commits you to actually focusing, rather than quietly drifting back into trying to do everything.

Build the governance machinery

The governance chapter described what to build. Now you build it. The pre-launch checklist. The risk classification process. The AI risk officer role, if it does not exist. The recurring audit calendar. The model card template. The incident response playbook. The shadow-AI inventory and approval process.

None of this is glamorous and all of it is important. The single most useful thing you can do in days thirty-one to sixty is build the system that will let you say no to bad ideas and yes to good ones with the same authority. Without it, you will spend the next two years adjudicating each new AI proposal by the force of your personality, which is exhausting and does not scale.

Do not try to build the perfect governance system in month two. Build the minimum viable version, ship it, and improve it as you learn. The companies that try to design comprehensive governance before deploying any of it end up with documents nobody uses.

Have the hard conversations

Month two is when you have the difficult conversations the previous regime avoided. The project that is consuming engineering time but has no realistic path to value. The vendor contract that is overpriced and over-promised. The team that is structured wrong for the work it now needs to do. The senior engineer who is technically excellent but is blocking the broader team's progress. The compliance gap that everyone has known about and nobody has fixed.

These conversations are why you were hired. The previous leader either could not or did not have them. Do not delay. The longer you wait, the more the organisation will assume you are not going to have them at all, and the harder they become.

Be specific. Be respectful. Be willing to be wrong about any individual decision while being clear that decisions are now being made. The conversations will be hard in the moment and useful for the next two years.

Days 61 to 90: Ship and learn

By day sixty you have a focused set of priorities, a working governance system, and a clear-eyed view of what your organisation can actually deliver. Now you produce visible results.

Three things to do in the final thirty days.

Ship the first thing

Pick the most important item on your focus list. Move it from planning to production in the next thirty days. This will not be the biggest project. It will be the project where the team is ready, the requirements are clear, and the path is short. Concentrate resources on it. Help unblock it personally. Make sure it ships.

The first ship matters more than its specific success. It establishes that under your leadership, things move. It rebuilds the team's confidence in their ability to deliver. It gives the CEO and the board something to point to. It is your first proof point. Do not let it slip past day ninety.

Establish the rituals

Set up the recurring meetings, reviews, and audits that will run after your direct attention has moved elsewhere. The weekly leadership check-in on the focus list. The monthly model audit. The quarterly board update on AI progress and risk. The semi-annual review of which AI systems still earn their cost.

None of these are exciting. All of them are what turns a personal effort into an institutional one. If you leave the company in two years and these rituals continue without you, you have built something durable. If they stop the day you leave, you have only ever managed by personal force, which is fragile.

Tell the story

In the final week of the ninety days, communicate broadly what you have learned, what you have changed, and what is next. To the engineering team. To the broader company. To the board. To customers, in the cases where it is appropriate.

The story should be specific. Not "we are taking AI seriously," which means nothing. Specific: "In the last ninety days we inventoried our AI systems, reduced our active AI projects from twenty-one to four, established governance with the authority to halt launches, shipped our first product feature under the new system, and have committed to the following measurable outcomes for the next quarter."

The specificity protects you. Vague communication is what allows perception to drift from reality. Concrete communication is what builds the kind of trust that survives the inevitable bad weeks that come after the honeymoon.

After day 90

The ninety days are not the job. They are the foundation for the job. If you have done them well, you now have:

- A current inventory of every AI system, classified by risk.
- A focused set of priorities the CEO and the board can defend.
- A governance system that lets you say no with authority.
- A trusted relationship with thirty key people across the organisation.
- A first visible shipment under the new approach.
- A set of recurring rituals that will outlast your direct attention.

From here, the work is to keep doing the work. Update the inventory. Sharpen the focus. Strengthen the governance. Hire the team. Ship the next thing. Defend the priorities when they come under pressure. Refresh your understanding of the technology as it keeps changing underneath you. None of this is dramatic. All of it is the job.

A composite case

A retail company hired their first head of AI in early 2024. She was an experienced data scientist who had led teams at two larger companies. The CEO wanted her to move fast. The board wanted to see momentum within the quarter.

She did not move fast in the first thirty days. She did the work I just described. She inventoried thirty-one AI systems, of which the official list had only nineteen. She met fifty-two people, against the thirty I would have advised, because the company was bigger than she had expected. She read every document she could find, and discovered three projects that had been quietly funded out of separate budgets without her predecessor's knowledge. She sat with the customer service logs for two full days and discovered that the AI summarisation tool the company was proud of was systematically misrepresenting customer complaints about a specific product line, which the product team had not yet figured out.

In month two she made the hard cuts. The twenty-one active AI projects became four. She moved the AI risk officer role out of engineering and into legal, against the objections of the CTO, on the grounds that the previous structure had no real independence. She had three hard conversations with senior engineers about projects she was stopping, and lost one of them to a competitor within the month, which she accepted as a cost of the focus.

In month three she shipped a new version of the customer service summarisation tool, retrained with explicit attention to the failure mode she had found in month one. The new version was measurably more accurate, and the product team caught the issue with their problem product line within a week of the launch because the summaries were finally telling them what customers were actually saying.

At the end of ninety days she presented her work to the board. She said, in her own words, "I spent the first month understanding what we had, the second month deciding what we were going to do, and the third month doing the first piece of it. The next ninety days will produce more. The pace will accelerate from here."

She is still in that role two years later. The AI program at the company is one of the more thoughtful programs I have seen at any company of comparable size. When I asked her later what had made the difference, she said it was the discipline of not announcing anything in month one. "I could feel the pressure to perform," she told me. "I chose to wait. Everything else was downstream of that."

Questions for your first ninety days

To ask yourself, honestly, on day 30, day 60, and day 90.

94. On day 30: Do I know more about this organisation's AI than I did when I started, and do

my notes reflect a specific picture rather than a general one?

95. On day 30: Have I avoided announcing anything I will regret?
96. On day 60: Are my priorities written down, defensible, and shared with the people who need to support them?
97. On day 60: Have I had the difficult conversations I was hired to have, or have I quietly postponed them?
98. On day 90: Have I shipped something visible under the new approach?
99. On day 90: If I left tomorrow, would the work survive me, or does it all depend on my personal attention?

SIX THINGS TO TAKE FROM THIS CHAPTER

The first 30 days are for listening. Do not announce, restructure, or commit. Build the true picture.

The second 30 days are for focus and governance. Pick the three to five things that matter. Build the machinery that lets you say no.

The third 30 days are for shipping and ritual. Land the first visible win. Establish the cadence that outlasts you.

Most failed AI leadership tenures fail in month one, because the leader acted before they knew what was true.

The pressure to demonstrate momentum is the enemy of doing the foundational work that makes momentum sustainable.

If your ninety days produce a clear picture, a defensible focus, working governance, and one visible win, you have done the job. Everything after that is leverage on the foundation.

ONE-LINE TAKEAWAY

Listen first. Focus second. Ship third. The order is not interchangeable. The leaders who skip the listening produce noise. The leaders who skip the focus produce sprawl. The leaders who skip the shipping produce strategy decks. The leaders who do all three, in sequence, produce a foundation that supports the next decade of work.

CHAPTER A

Glossary

Plain definitions for the terms used in this book and in most conversations about data and AI. When a definition is contested in the field, I have given the most useful working version, not the most academic one.

Accuracy
The fraction of a model's predictions that are correct. A coarse metric that hides important details (see precision and recall).

Adversarial training
Deliberately training a model on inputs designed to fool it, so it becomes more robust to attack.

Algorithm
A precisely specified procedure for solving a problem. Machine learning algorithms learn the procedure from data; classical algorithms have the procedure written by humans.

Bias (statistical)
A systematic error in a model's predictions, usually rooted in skewed training data.

Bias (fairness)
The same problem, viewed through the lens of who is helped or harmed by the systematic error.

Cloud computing
Renting computing resources from a third party that operates them in their own data centres.

Data engineer
A specialist who builds the systems that move and clean data.

Data lake
A storage system that holds raw, often unstructured data at scale, to be processed later.

Data scientist
A specialist who analyses data to find patterns and builds predictive models.

Data warehouse
A structured storage system optimised for analytical queries on cleaned, integrated data.

Deep learning
A subfield of machine learning that uses neural networks with many layers. Dominant in current research and product development.

Defense in depth
A security strategy that uses multiple overlapping layers, assuming any single layer will eventually fail.

ELT (Extract, Load, Transform)
The modern data pipeline pattern: raw data is loaded into the warehouse first and transformed there.

ETL (Extract, Transform, Load)
The classical pattern: data is cleaned in flight before being loaded into the warehouse.

Explainable AI (XAI)
Techniques that allow humans to understand why a model produced a given prediction.

Feature
A single input variable used by a model.

Feature store
A central system that stores engineered features for reuse across models and between training and serving.

IaaS (Infrastructure as a Service)
A cloud model where the provider gives raw computing resources and the customer manages everything above.

Inference
Running a trained model on new data to produce predictions.

Large language model (LLM)
A neural network trained on enormous text corpora to predict language patterns. The technology behind today's leading conversational AI assistants from the major model labs.

Machine learning
A subfield of AI in which systems learn from data rather than from hand-written rules.

MLOps
The practice of running machine learning systems reliably in production, borrowing heavily from DevOps.

Model
The trained mathematical artefact that takes inputs and produces predictions.

Model drift
Degradation of a model's accuracy over time because the world has changed since the model was trained.

Multi-factor authentication (MFA)
An access control practice that requires more than one verification factor, typically something the user knows plus something they have.

Overfitting
When a model memorises its training data instead of learning generalisable patterns.

PaaS (Platform as a Service)
A cloud model where the provider also handles the operating system and common services.

Pipeline
A connected series of steps that processes data from one form to another.

Precision
Of the items a model said yes to, the fraction that were actually yes.

Predictive maintenance
Using sensor data and machine learning to predict equipment failure before it happens.

Recall
Of the items that were actually yes, the fraction the model correctly identified.

Reinforcement learning
A learning paradigm where an agent learns by taking actions in an environment and receiving rewards.

Role-based access control (RBAC)
An access control model that assigns permissions to roles and roles to users.

SaaS (Software as a Service)
A cloud model where the provider runs the entire application and the customer simply uses it.

Supervised learning
Learning from data where each example has a correct answer attached.

Training
The process of fitting a model to data.

Transformer
The neural network architecture, introduced in 2017, that underlies almost all modern large language models.

Unsupervised learning
Learning to find structure in data without labelled answers.

Agent
An LLM placed inside a loop with access to tools, allowed to take multiple steps to accomplish a task. An LLM with hands.

Agentic workflow
A multi-step task carried out by an agent, where each step's output informs the next step's input.

Chain of thought
A prompting technique where the model is asked to show its reasoning step by step before producing a final answer. Sometimes improves accuracy on complex tasks.

Context window
The maximum amount of text an LLM can consider in a single query. Larger windows allow longer documents or conversations as input.

Embedding
A representation of text (or other data) as a list of numbers, such that semantically similar items have similar numbers. The foundation of modern semantic search.

Evaluation harness
A system that runs an LLM application against a curated set of test cases and measures the quality of the output, used to detect regressions when prompts or models change.

EU AI Act
The European Union's comprehensive AI regulation, passed in 2024, with phased enforcement through 2027. Sets the de-facto global standard for AI compliance.

Fine-tuning
Training a pre-existing model further on your specific data to specialise its behaviour. The third option after prompting and RAG; reach for it only when the first two fall short.

Frontier model
The most capable LLMs available from major labs (OpenAI, Anthropic, Google). Most expensive per query, most general in capability.

Function calling
An LLM's ability to invoke external tools or APIs as part of generating its response, by producing structured calls the application then executes.

Hallucination
When an LLM confidently produces output that is factually wrong. A core failure mode of the technology, mitigated but not eliminated.

High-risk AI system
An EU AI Act classification for AI used in employment, education, credit, healthcare, law enforcement, migration, justice, and operation of critical infrastructure. Heavy compliance burden.

Hybrid pricing
A pricing model combining a flat fee with usage-based charges above a threshold. The model most companies eventually adopt for AI features.

Inference
Running a trained model on new data to produce predictions. The dominant cost of AI in production.

LLM (large language model)
A neural network trained on enormous text corpora to predict language patterns. The technology behind today's leading conversational AI assistants.

LLMOps
The discipline of running LLM applications reliably in production. Borrows from MLOps but adds LLM-specific concerns like prompt versioning and evaluation harnesses.

Model card
A standardised document describing a model's intended use, training data, known limitations, performance metrics, and the conditions under which it was tested.

Model registry
A central system for storing trained models with their version, metrics, and metadata. The answer to "which model is actually running in production?"

Open weights model
An LLM whose weights are publicly available, allowing organisations to run the model on their own infrastructure. Llama and Mistral are examples.

Prompt
The input given to an LLM to produce a response. The discipline of writing effective prompts is non-trivial.

Prompt engineering
The practice of crafting prompts to reliably produce desired outputs from an LLM. A skill, not a science.

Prompt injection
An attack in which an adversary embeds instructions in user input that an LLM-based system inadvertently follows. The most common new vulnerability of the LLM era.

RAG (retrieval-augmented generation)
An architectural pattern where relevant private documents are retrieved at query time and provided to the LLM along with the user's question. The dominant pattern for enterprise LLM applications.

Responsible AI
The procedural and governance side of AI safety. Who decides, who oversees, who is accountable. Distinct from ethical AI.

Ethical AI
The substantive question of what an AI system does to the people it affects, and whether that is right. Distinct from responsible AI.

Shadow AI
AI tools being used inside an organisation without formal approval, often through SaaS features that came bundled with other software. Almost always larger than the inventoried set.

SHAP (SHapley Additive exPlanations)
A technique for explaining individual predictions of a machine learning model in terms of how each input feature contributed.

Specialised small model
A smaller, focused model trained or fine-tuned for a narrow task. Often cheaper, faster, and more controllable than a frontier model for a bounded use case.

System card
Similar to a model card but describing the entire system around the model, including the surrounding controls and intended deployment context.

System prompt
The instructions given to an LLM that set its role, tone, and behavioural rules, separate from the user's individual queries.

Token
The unit in which LLMs process text, roughly equivalent to a short word or word fragment. Both context windows and pricing are measured in tokens.

Vector database
A database optimised for storing embeddings and finding similar ones quickly. Pinecone, Weaviate, pgvector are examples.

Vibe coding
The informal name for the practice of describing what you want in natural language and letting an AI assistant generate code, which the engineer then reviews and refines.

CHAPTER B

Worksheets for Leaders

Six tools designed to make this book usable in a working week. Print them. Fill them in. Bring them to your next leadership meeting.

Worksheet 1, AI Readiness Assessment

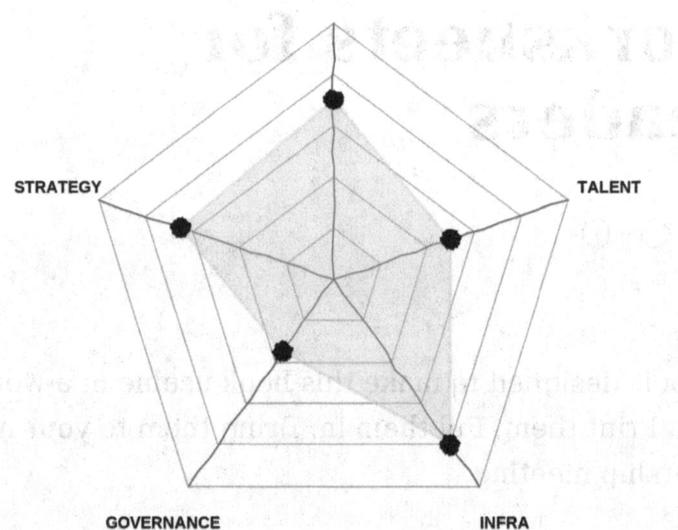

Score your organisation on each of the five dimensions from 1 (just starting) to 5 (mature). Draw the result on the radar above.

Data: _____ (have we got it, is it clean, can we trust it)

Talent: _____ (do we have or can we recruit the team)

Infrastructure: _____ (is the cloud and tooling foundation in place)

Governance: _____ (are policies, ethics, and security frameworks in place)

Strategy: _____ (do we know what we are trying to achieve with AI)

Lowest score: __________ That is the focus for the next six months.

Worksheet 2, Build vs Buy vs Partner

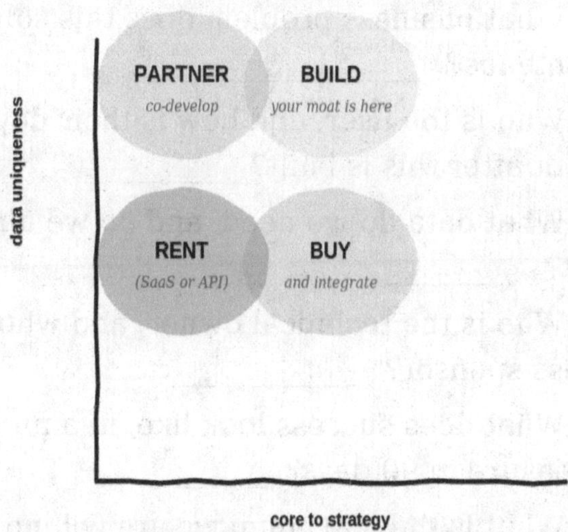

For the AI capability you are considering, plot it on the matrix above.

Capability under consideration:

____

Is this data uniquely ours? YES / NO (circle)

Is this core to our strategy? YES / NO (circle)

Decision (build / buy / partner / rent):

____

If we are wrong about this, what is the cost of reversal?

____

____

Worksheet 3, The AI Pilot Checklist

Before approving any AI pilot, answer these six questions in writing.

100. What business problem does this solve, in one sentence? ____________________
101. Who is the user, and how is their day different after this is built? __________
102. What data do we need, and do we already have it? ____________________________
103. Who is the technical owner, and who is the business sponsor? ______________
104. What does success look like, in a number we can measure in 90 days? ________
105. What is the maximum we are willing to spend before pulling the plug? _______

Worksheet 4, Cybersecurity Posture Check

Honest answers only. Use the scores to brief your CISO.

Do all employees use multi-factor authentication? YES / NO / NOT SURE

Do we have a complete inventory of systems and access? YES / NO / NOT SURE

Are access grants reviewed quarterly? YES / NO / NOT SURE

Do we have written incident response procedures? YES / NO / NOT SURE

Has our security team trained on AI-specific threats? YES / NO / NOT SURE

Number of NO or NOT SURE answers: _____

Any number above zero is a top-three priority for this quarter.

Worksheet 5, The Ethical AI Rubric

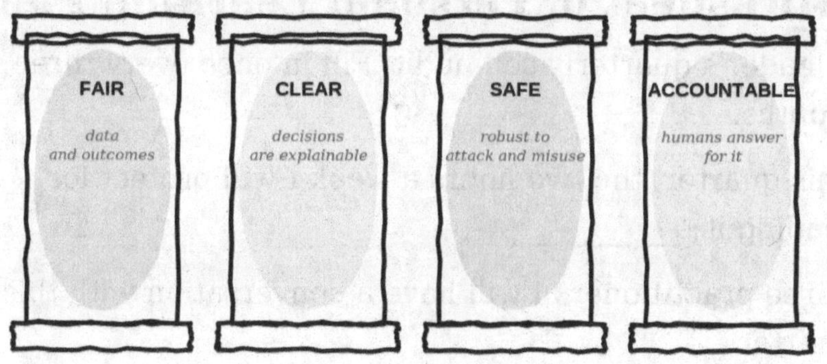

your AI system stands on all four, or none

For every AI system you deploy, answer four questions.

FAIR: Could this model systematically disadvantage any group?

____

____

CLEAR: Can we explain a specific decision to the person affected by it?

____

____

SAFE: Is the model robust to adversarial inputs and data poisoning?

____

____

ACCOUNTABLE: Who, by name, is responsible when this model is wrong?

____

____

Worksheet 6, Personal Learning Plan

A leader's quarterly self-audit. Fill in once every three months.

This quarter, the two hours a week I will protect for reading is: ____________

Three practitioners I will have a conversation with this quarter:

1. ____

____

2. ____

____

3. ____

____

One small thing I will build or try myself:

____

What I will tell my team I have learned that they should pay attention to:

____

____

Worksheet 7, The AI Pilot Scorecard

Complete before the pilot begins. Sign off by the executive sponsor. Do not change after the pilot starts. This is the single most useful piece of paper in the entire appendix.

Pilot name:

____

_

Executive sponsor (by name):

____

The hypothesis being tested (in one sentence):

____

____

____

____

To scale, the pilot must achieve, by week 12:

Primary metric: _______________________ Target:

____

Secondary metric: ____________________ Floor (must not drop below): ____________

To kill, the pilot need only fail to:

____

____

Total pilot budget (all in): $___________ Duration: ___ weeks

Sponsor signature: ________________ Date: ________

Worksheet 8, Shadow AI Audit

To be filled in after a department-by-department conversation. The first audit will surprise you. The second audit, three months later, will surprise you again.

Department:

____

_

AI tools currently in use (named):

1. ______________________ Purpose: _______________ Approved: Y/N

2. ______________________ Purpose: _______________ Approved: Y/N

3. ______________________ Purpose: _______________ Approved: Y/N

4. ______________________ Purpose: _______________ Approved: Y/N

5. ______________________ Purpose: _______________ Approved: Y/N

What customer or employee data passes through these tools? Be specific.

____

____

Tools to ban:

____

_

Tools to approve formally with controls:

____

Tools to replace with an approved alternative:

____

Worksheet 9, The Vendor Pitch Decoder

Before the vendor's next pitch meeting. After the meeting, score yourself honestly.

Vendor: ____________________ Date of pitch:

____

Five questions to ask the vendor in the meeting:

106. Show me the model card. What data was the model trained on? What are its known limitations?
107. Describe a recent failure of your product, and what you did about it. (Vendors who claim no failures are not credible.)
108. What happens to our data when it passes through your system? Where is it stored? Who has access?
109. How do you measure quality of output? Show me an actual evaluation against our use case.

110. If we end the contract in two years, how do we get our data out, and how long does that take?

Red flags noted during the pitch:

Used "AI," "ML," "deep learning" interchangeably: Y / N

Could not name a specific failure mode of their product: Y / N

Demo was on their data, not ours: Y / N

Pricing depended on "call us": Y / N

Three or more Yes answers means run another pilot before committing.

Worksheet 10, The First-90-Days Checklist

For your first quarter as an AI leader. Tick as you go. Be honest.

Day 30:

[ ] Complete inventory of all AI systems (including shadow AI) compiled.

[ ] One-on-one conversations completed with at least 30 key people.

[ ] All existing AI documentation read and notes captured.

[ ] Production logs reviewed for at least one full day.

[ ] No public commitments made about strategy yet.

Day 60:

[ ] Focus list of 3 to 5 priorities written and shared.

[ ] Projects outside the focus paused, killed, or maintained at minimum.

[ ] Governance machinery built (pre-launch checklist, risk classification, incident response).

[ ] AI risk officer role staffed (or escalated to the CEO if not).

[ ] The hard conversations have happened, not been delayed.

Day 90:

[ ] First visible launch under the new approach has shipped.

[ ] Recurring rituals (audits, reviews) established and on the calendar.

[ ] Story communicated to engineering, the broader company, the board.

[ ] Plan for the next 90 days written.

CHAPTER C

Real-World Case Studies

Public examples from the past decade, drawn from sources cited at the end of each study. I have deliberately mixed wins and failures. The failures teach more.

The Netflix Prize (2009)

Won by a team that combined many models into an ensemble that beat Netflix's own recommender by ten percent. Netflix never deployed the winning algorithm in production because the engineering cost of running it exceeded the additional value it would produce. A foundational lesson in the gap between research-grade and production-grade AI. (Source: Netflix Tech Blog, 2012)

IBM Watson in Oncology (2013, 2018)

IBM partnered with MD Anderson Cancer Center to apply Watson to cancer treatment recommendations. The project was abandoned in 2017 after producing recommendations that physicians found unreliable. The case became a touchstone for the gap between AI marketing and clinical reality. (Source: STAT News investigation, 2017)

Amazon's Recruiting Tool (2018)

Amazon built an AI tool to screen resumes. It was abandoned after the team discovered it systematically downgraded resumes containing the word "women's" (as in "women's chess club"). The training data, ten years of resumes from a male-dominated industry, had taught the model that male candidates were preferred. A landmark case in how bias enters models through historical data. (Source: Reuters, 2018)

AlphaFold (2020, 2021)

DeepMind's protein structure prediction system solved a problem biologists had worked on for fifty years. The model and its predictions for over two hundred million proteins were released free for research use. One of the clearest cases of AI delivering genuine scientific value. (Source: Nature, 2021)

GitHub Copilot (2021)

An AI pair-programming assistant trained on public code. Widely adopted by developers, also the subject of ongoing legal disputes about training on open-source code without licence compliance. A case study in deploying useful AI before the legal frameworks catch up. (Source: GitHub, ongoing)

The Apple Card Credit Limit Controversy (2019)

Customers reported that the Apple Card, underwritten by Goldman Sachs, offered women lower credit limits than men with similar profiles, including spouses. Investigation found no intentional discrimination, but the model's outputs were difficult to explain even to Apple's executives. A case study in why explainability matters operationally, not just ethically. (Source: New York State Department of Financial Services investigation, 2021)

Tesla Autopilot's Long Tail (2014 to present)
Tesla's driver assistance system has performed remarkably well in most conditions and disastrously in a small number of edge cases. The long tail of failure modes in autonomous systems is the single hardest engineering problem in production AI. (Source: NHTSA investigations, ongoing)

The Consumer LLM Breakthrough (2022)
OpenAI's late-2022 release of a flagship conversational large language model moved the category from research curiosity to mainstream tool in approximately three months. The launch is now studied as a case in how rapidly a technology category can shift when a usability barrier is removed. (Source: widely reported)

CHAPTER D

Resources by Chapter

Opinionated. I have read or used everything here.

Chapter 1, How We Got Here

- Book: "The Quest for Artificial Intelligence" by Nils J. Nilsson. A definitive academic history, free PDF available from Stanford.
- Essay: "The Bitter Lesson" by Rich Sutton, 2019. A short, important argument about why scale beats cleverness.

Chapter 2, The Basics, Plainly

- Book: "Hands-On Machine Learning" by Aurélien Géron. The best practitioner-level introduction.
- Book: "The Hundred-Page Machine Learning Book" by Andriy Burkov. The best leader-level introduction.

Chapter 3, Where the Cloud Fits

- Documentation: AWS, Azure, and Google Cloud all have well-written architecture frameworks. Pick the one for your primary cloud.

- Newsletter: "The Pragmatic Engineer" by Gergely Orosz. Current, practical, vendor-neutral.

Chapter 4, Architecting AI That Doesn't Break

- Book: "Designing Machine Learning Systems" by Chip Huyen. The standard reference for MLOps.
- Book: "Reliable Machine Learning" by Cathy Chen et al. Strong on the operational realities.

Chapter 5, Securing What You Build

- Book: "Cybersecurity and Infrastructure Security Agency (CISA) publications", freely available.
- Book: "Adversarial Machine Learning" by Anthony D. Joseph et al. The reference on AI-specific attacks.

Chapter 6, The Team Behind the Tech

- Book: "The Manager's Path" by Camille Fournier. Best book on managing technical people.
- Book: "An Elegant Puzzle" by Will Larson. Best book on managing engineering organisations.

Chapter 7, Data Engineering, the Discipline

- Book: "Fundamentals of Data Engineering" by Joe Reis and Matt Housley. The current standard reference.
- Newsletter: "Data Engineering Weekly" by Ananth Packkildurai.

Chapter 8, Staying Useful

- Newsletter: "Stratechery" by Ben Thompson. The single most useful technology newsletter for leaders.

- Practice: One serious technical conversation per quarter with a working practitioner. Buy them lunch.

Chapter 9, What an LLM Actually Is

- Book: "On the Origin of Good Moves" by Cole Wagner (an idiosyncratic and useful introduction to how large models actually learn).
- Article: Stephen Wolfram's plain-language explanation of how large language models work, freely available on his blog. The clearest non-technical explanation in print.
- Practice: Spend an hour with a frontier LLM each week, asking it questions in your domain. The intuitions you build are not transferable from reading about it.

Chapter 10, Agents Are Not a Miracle

- Paper: "ReAct: Synergizing Reasoning and Acting in Language Models" by Yao et al., 2022. The foundational paper on agentic patterns. Worth reading even if you do not normally read papers.
- Tool: Build a small agent yourself with an open framework before you make any vendor decision. Two hours of personal experience beats two weeks of vendor pitches.
- Practice: Audit your own agent deployments quarterly. Treat each agent as a junior employee in performance review.

Chapter 11, The Generative AI Stack

- Book: "Designing Data-Intensive Applications" by Martin Kleppmann. Predates the LLM stack but the principles transfer directly.
- Documentation: Read your chosen vector database's documentation cover to cover. It is shorter than you expect and dramatically improves the conversations you can have with your engineering team.
- Practice: Re-evaluate every component of your stack every six months. The category is moving too fast to commit to anything indefinitely.

Chapter 12, What the Machine Cannot Do (AI-Assisted Development)

- Newsletter: "The Pragmatic Engineer" by Gergely Orosz. Best current coverage of how engineering organisations are actually changing.
- Practice: Sit with one of your senior engineers for a full hour as they work with their AI coding assistant. Do not interrupt. The pattern you observe is more useful than any report.
- Practice: Once a quarter, have a one-on-one with a junior engineer about how they are learning. The pipeline problem only becomes visible if you ask.

Chapter 13, The Letter You Do Not Want to Receive (Governance)

- Document: The full text of the EU AI Act. Long, dense, useful. Have at least one senior leader who has actually read it.
- Newsletter: "Tech Policy Press" for current AI regulatory developments globally.

- Practice: A quarterly tabletop exercise on an AI compliance scenario. Find out where your gaps are before a regulator does.

Chapter 14, Ethics and Responsibility Are Not the Same Thing

- Book: "Weapons of Math Destruction" by Cathy O'Neil. Older but still the clearest articulation of how algorithmic systems can do harm at scale.
- Book: "The Alignment Problem" by Brian Christian. The intellectual history of the field, accessibly written.
- Practice: Once a quarter, look at your AI systems' outputs broken down by demographic group. Be willing to act on what you find.

Chapter 15, The Bill That Surprised the CFO (Economics)

- Tool: Your cloud provider's cost calculator. Run your AI workload through it at three volumes (current, 2x, 10x) and look at the curve.
- Practice: Re-cost every major AI use case every six months. The cost curve is moving faster than your planning horizon.
- Conversation: Have your CFO sit in on one AI architecture review per quarter. The vocabulary needs to go in both directions.

Chapter 16, Your First Thirty, Sixty, and Ninety Days

- Book: "The First 90 Days" by Michael Watkins. Not AI-specific, but the framework adapts.

- Practice: Write your day-30, day-60, and day-90 self-assessment in advance, as a contract with yourself. Re-read it on those days.
- Conversation: Find someone who has done this role before, anywhere. Buy them dinner. The lessons compound.

CHAPTER E

Implementation Playbooks

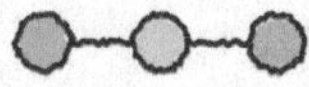

Six playbooks I have used or watched used at companies going through the most common AI transitions. None of them are the only way to do the work. All of them are how a competent leader, given the situation described, would approach it. Use them as a starting point for your own version, not as a recipe.

Each playbook is structured the same way: when you need it, what to do before you start, the phases of the work, what tends to go wrong, and what success looks like. If the format ends up feeling formulaic, blame me, not the work. The actual work in each case is harder and less tidy than any playbook can capture.

Playbook 1: Migrating Off Legacy Data Systems

When you need this: you have data trapped in systems that are decades old. A mainframe. A legacy ERP. A SQL Server instance that has been forgotten in a corner of a data centre for fifteen years. The business depends on the data. The systems depend on no one alive understanding them. You want to move.

Before you start

Identify the business value of the migration in measured terms. Not "modernisation," which means nothing. Specific things, like "reduce reporting cycle time from three weeks to one day" or "enable real-time fraud detection that the current system cannot support." If you cannot name the value, you are not ready to start. Migrations without measurable value tend to die in year two when budgets tighten.

Find the people, by name, who understand the existing systems. There are usually two or three, often near retirement, who hold critical knowledge in their heads. Plan to spend time with them before you start. Plan to pay them well, or pay external consultants who know systems like theirs, to document what they know before they leave.

Phases

Phase 1, months 1 to 3: Inventory and understand. Map the existing systems. Document the data flows, the consumers, the dependencies you do not yet know about. Build the target architecture at a high level. Get sign-off from the business owners on what the migration will and will not change.

Phase 2, months 4 to 9: Pilot. Pick the smallest meaningful workload that uses the data, build the new version in the new system, run both in parallel, compare outputs. Fix the gaps you find. This phase tells you whether the migration is going to work.

Phase 3, months 10 to 15: Migrate. Move workloads in order from least risky to most risky. Each workload runs in parallel on old and new until you trust the new. Cut over only when the trust is established.

Phase 4, months 16 to 18: Decommission. Turn off the old systems carefully, in a documented sequence, with rollback plans for each step. Most companies skip this phase. The old systems then run forever, consuming budget and creating drift between official and actual sources of truth. Do the decommission deliberately.

What tends to go wrong

The business signs off on a scope that turns out to include data flows nobody mapped. The migration gets to month nine and discovers a critical downstream system that nobody told the team about. The fix is the inventory phase. Spend more time there than feels comfortable.

The pilot succeeds and the team assumes the migration will succeed. The real migration encounters scale, performance, and edge-case problems the pilot did not. Plan for at least three months of unexpected work after the pilot is declared a success.

The old systems never get decommissioned because the political conversation is too hard. Insist on it. The unfinished migration costs more than the finished one.

What success looks like

Eighteen months in, the old systems are off, the new ones are running, the measured business value is being captured, the people who knew the old systems have moved into new roles or retired with grace. The team that did the work understands the data layer of the business better than anyone alive ever has. That understanding is itself a durable asset.

Playbook 2: Your First Production ML Deployment

When you need this: your data science team has a model that works in a notebook. The business wants to use it. The team has never put a model into production. This is the moment most AI projects either become real or quietly die.

Before you start

Be honest about whether the model is actually ready. A model that achieves good metrics on a held-out test set is not necessarily ready for production. Ask, can the team explain confidently what the model does when the input looks different from training data? Can they describe how the model will be retrained as the data drifts? If the answers are vague, the model needs another month of work before you start this playbook.

Identify the human or system that will consume the model's output. Not in the abstract. Specifically. Are nurses going to read the predictions? Is a downstream system going to take action automatically? Are managers going to use the output to make decisions? The design of the deployment depends on the answer.

Phases

Phase 1, weeks 1 to 3: Productionise the code. The notebook becomes a service with an API. Tests are written. Logging is added. The model is loaded from a model registry, not from a hardcoded file path. The deployment is reproducible. None of this changes what the model does. All of it makes the model maintainable by people other than the one engineer who built it.

Phase 2, weeks 4 to 6: Shadow mode. The model runs in production but does not affect any decision. Its predictions are logged alongside the existing process. The team compares the model's predictions to actual outcomes for at least four weeks. This is where you discover that the model behaves differently in production than in testing, which it almost always does.

Phase 3, weeks 7 to 10: Limited deployment. The model affects a small fraction of decisions, in a controlled way, with humans able to override. Measure the impact on the business metric you said you cared about. Compare to the baseline. Be willing to roll back if the impact is negative.

Phase 4, weeks 11 onwards: Scale or kill. If the limited deployment is producing the value you expected, scale to the full population. If it is not, kill the project and learn from it. Do not let a project that is not working drift into wider deployment because someone is reluctant to pull the plug.

What tends to go wrong

Shadow mode reveals that the data the model sees in production is meaningfully different from the data it was trained on. The model's accuracy is lower than expected. The team is disappointed. The right response is to retrain the model on production-like data, not to ship anyway and hope it works out.

The team skips the shadow mode phase because the model looks good in testing. The model then fails in production in ways nobody predicted. The shadow mode phase exists for exactly this reason. Do not skip it.

Monitoring is added as an afterthought. Six months later the model has drifted and nobody notices because nobody set up the alerts. Build monitoring before you deploy, not after.

What success looks like

Three months in, the model is in production for the full population, measurably improving the business metric, with a working monitoring system that would alert the team if accuracy degraded. The team has documented enough about the deployment that a new engineer could maintain it. The next model takes less time to deploy because the infrastructure now exists.

Playbook 3: Building the Data Team from Zero

When you need this: your company has decided to invest seriously in data and AI. You have approval for headcount. You have no existing team. You are about to make hiring decisions that will shape the program for the next five years.

Before you start

Find an executive sponsor. Not a CEO who said "we should do AI." An executive who is willing to spend personal political capital protecting the team during the eighteen months it takes to deliver anything visible. Without this person, the team gets cut in the first budget pressure. With this person, the team gets the runway it needs.

Decide which part of the business the team will serve first. Not all of it. One area, deeply. Sales analytics. Or supply chain optimisation. Or customer support intelligence. The choice matters less than the focus. A team serving everyone at the start serves no one well.

Phases

Phase 1, months 1 to 3: Hire the data engineer first. This will feel wrong because everyone wants a data scientist. Do it anyway. The data engineer builds the pipelines that make later hires productive. Without them, the next data scientist you hire will spend a year being a data engineer and then leave.

Phase 2, months 4 to 6: Hire an analyst. Someone who can take questions from the business, run the queries, and produce useful answers. The analyst is the team's connection to the business. They establish credibility by shipping useful reports in the first quarter.

Phase 3, months 7 to 12: Hire the first data scientist. By now the pipelines work, the business is asking real questions, and the data scientist arrives into a productive environment. They can ship a first model in the first six months instead of spending that time cleaning data.

Phase 4, months 13 to 18: Specialise. Add an ML engineer to take models to production. Add a second analyst to cover more business areas. Add a senior data scientist who can mentor the first one. The team starts to look like a real organisation.

What tends to go wrong

The hiring order gets reversed. A data scientist is hired first, spends a year being a data engineer, leaves. The team has to start over. The hiring order is the most counter-intuitive part of this playbook and the most important.

The team takes on too many requests too early. Saying yes to everyone makes them ineffective for anyone. The executive sponsor's first job is to help the team say no in the early months.

The team is hired but no one defends them when budgets tighten. They get cut before they have shipped anything visible. The eighteen-month commitment from the sponsor is what protects against this.

What success looks like

Eighteen months in, the team has five to seven people, has shipped two or three models that the business actually uses, is producing recurring analytics that show up in executive reviews, and is hiring its second wave from a position of demonstrated value. The CEO no longer asks "what are we getting from the data team." They ask "how can we get more of this."

Playbook 4: Responding to Your First AI Security Incident

When you need this: something has gone wrong. Your model leaked data. Your AI was attacked. Your chatbot said something harmful that has hit the press. Your model made a decision that has triggered a regulator's letter. The clock is now running and most of your future will be decided by how you respond in the next 72 hours.

Before you start

You should have built the incident response playbook before the incident, not after. If you are reading this during a live incident, follow it anyway. If you are reading this in advance, rehearse the playbook with your team quarterly so the muscle is there when you need it.

Phases

Phase 1, hour 0 to 6: Contain. Stop the bleeding. If a model is producing harmful output, take it offline. If data is leaking, cut the access. If an attack is in progress, isolate the affected systems. Do not wait for full understanding before acting. Containment buys you time to think.

Phase 2, hour 6 to 24: Assemble the response team. Engineering, security, legal, communications, the executive who will own the public-facing response. Get them in one room or one video call. Establish a single source of truth for what is known. Avoid the trap of fifty Slack channels with conflicting fragments of information.

Phase 3, day 2 to 5: Investigate. What happened, when, how, what was affected. Preserve evidence. Document the timeline. Do not destroy logs or rotate compromised credentials without first capturing what you need to capture. Most companies' first instinct is to clean up immediately. Investigators need the mess intact long enough to understand it.

Phase 4, day 2 onwards (in parallel): Communicate. Affected customers first. Regulators where required (and the requirements often have deadlines measured in hours, not days). Internal employees who need to know. Public communications only when you have enough understanding to be truthful, and even then, brief and factual rather than reassuring.

Phase 5, week 2 to 6: Remediate. Fix the underlying problem. Strengthen the defences that failed. Update the playbook for next time. Communicate the fix to affected parties.

Phase 6, week 6 to 8: Learn. Hold a blameless post-mortem. Document what failed. Document what worked. Update training. Share the lessons widely inside the company. The companies that handle their second incident better than their first are the ones that took the first one seriously as a learning event.

What tends to go wrong

The team tries to handle the incident quietly without involving legal or communications. The story breaks before the company is ready, and the public version is now controlled by someone other than the company. Bring legal and communications in on day one.

The investigation is rushed and the root cause is misidentified. The fix solves a symptom but the underlying problem remains. The next incident, from the same root cause, is worse because the company already claimed to have fixed it.

The post-mortem becomes a blame exercise. People stop telling the truth. The lessons are lost. The post-mortem must be blameless or it does not work.

What success looks like

Six weeks in, the immediate harm is contained, affected parties have been treated fairly, the root cause has been fixed, the regulator (if involved) is satisfied with the response, and the company has documented lessons that will make the next incident easier to handle. Trust takes longer to rebuild. The incident response is the foundation for the trust-rebuilding work, not a substitute for it.

Playbook 5: Choosing Your First Cloud Provider

When you need this: you are either making a greenfield cloud decision or re-evaluating after a previous decision has not aged well. The choice will shape your infrastructure for the next five to ten years.

Before you start

Write down what you actually need, before any vendor talks to you. Specifically: what workloads, what scale, what compliance requirements, what existing skill in your team, what existing tools you must integrate with. This document is your shield against vendor demos.

Without it you will be talked into something based on the impressiveness of the pitch, not the fit with your needs.

Phases

Phase 1, weeks 1 to 3: Define needs. The document described above. Sign off internally. Do not start vendor conversations until it is done.

Phase 2, weeks 4 to 6: Shortlist. Reduce the choice to two or three providers. The big three (AWS, Azure, Google Cloud) cover most needs. Specialist providers like Oracle Cloud or IBM may be on the list for specific industry needs. Eliminate providers that do not meet your non-negotiables first, then evaluate the rest deeply.

Phase 3, weeks 7 to 12: Pilot. Run a representative workload on each shortlisted provider. Measure real performance, real cost, real developer experience. This phase is what separates the decision-by-vendor-pitch from the decision-by-evidence. The pilot is meaningful work, do not let any vendor talk you out of it.

Phase 4, week 13: Decide. Document the decision and the reasoning. Communicate to the organisation. Begin the migration plan.

What tends to go wrong

The decision is made for political reasons (the CEO knows a vendor, the CTO worked at a vendor) without honest technical evaluation. These decisions tend to be regretted within two years. Insist on the pilot phase regardless of who has a relationship with whom.

The pilot is too narrow and does not surface the issues that will matter at scale. Make the pilot workload as close as possible to your actual most-demanding workload. A pilot that runs a toy example tells you nothing about how the provider will perform when it matters.

The team commits to a provider, then quietly starts using a second one for specific projects, then a third. The multi-cloud creep begins. Have an explicit policy on when a second cloud is allowed and require executive approval for each exception.

What success looks like

Three months in, the decision is made with documented reasoning, the migration plan exists, the team is being trained on the chosen provider, and the executive team can defend the choice publicly. Five years later the decision has aged well enough that you do not have to make this choice again.

Playbook 6: Running an AI Pilot That Actually Proves Something

When you need this: you are considering a major AI investment and want to test the assumption before committing. This is the single most useful discipline in AI leadership and the one most often skipped. A pilot that proves something is worth ten that prove nothing.

Before you start

Write down, in advance, what would convince you to scale the project, and what would convince you to kill it. Specifically. Not "if it works, we will scale." Specifically: "if the model reduces support ticket resolution time by at least 30 percent for the chosen ticket category, and customer satisfaction does not drop by more than 2 points, we will scale. Otherwise we will kill it."

Get the success criteria signed off by the executive sponsor before the pilot starts. The reason for this is that sunk-cost reasoning is powerful. After three months of pilot work, the team will want to continue regardless of results. The pre-committed criteria are what protects you from continuing a project that should be killed.

Phases

Phase 1, week 1: Define success criteria in writing. As described above. Sign off by the sponsor. This document is the contract for the rest of the work.

Phase 2, weeks 2 to 3: Scope ruthlessly. Cut the pilot to the smallest meaningful test of the hypothesis. Not the smallest version of the eventual product. The smallest test of whether the hypothesis is true. These are different. The team will want to scope larger. Resist. The smaller the pilot, the faster the answer.

Phase 3, weeks 4 to 9: Run. Build the pilot. Run it on real data with real users where possible. Measure against the success criteria. Do not move the criteria during the pilot, no matter what.

Phase 4, weeks 10 to 11: Measure honestly. Compare results to the pre-committed criteria. Be willing to find that the criteria were not met.

Phase 5, week 12: Decide. Scale or kill, based on the documented criteria. Write up the lessons either way. A kill that produces clear lessons is a valuable outcome. A scale that produces no lessons is not.

What tends to go wrong

The criteria are vague and effectively unmeasurable. The team interprets ambiguous results as success. The project scales. It then fails at production scale and the company has lost a year. Sharper criteria catch this before the scale decision.

The team becomes attached to the pilot and resists killing it even when the criteria are not met. The executive sponsor's job is to enforce the pre-committed criteria. This is sometimes the hardest part of their job.

The pilot succeeds in a narrow sense but the success would not generalise. The pilot was on a small clean dataset; production data is messier. The pilot was for a known cooperative user; production users are not all cooperative. Choose pilot conditions that resemble production as closely as possible.

What success looks like

Twelve weeks in, you have a clear yes-or-no answer about whether the AI investment is worth scaling. Either way, the answer is defensible to the board. Either way, the team has learned something durable. The pilot that produced a no is just as valuable as the pilot that produced a yes, because both saved the company from a much more expensive mistake.

A final note on all six playbooks: they are not substitutes for thinking. They are scaffolding for the thinking you have to do anyway. Adapt them to your situation. Skip the steps that do not apply. Add steps that the playbooks miss. The playbooks are a starting point, not a constraint.

CHAPTER

About the Author

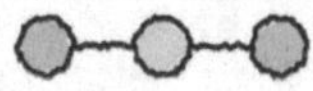

Sarah Choudhary is a technologist, founder, and writer. She grew up in Islamabad taking household appliances apart, studied computing as an undergraduate in Pakistan, and moved to the United States for graduate work. She holds a Masters and PhD in Data Science.

Over two decades she has worked across multiple industries on data and AI systems, led teams that have shipped products in production, and advised organisations through digital transformations. She believes the best leadership advice on technology is honest, specific, and short. She is allergic to acronyms that have stopped meaning anything.

She is the author of three books, of which this is the most technically substantial. The other two, Brew Your Business and Unleashed, focus on founders and the inner game of entrepreneurship respectively. They share a worldview, you can read them in any order.

She lives with her family and is forever grateful for their patience while she chased one more idea.

Want to work with Sarah?

- Newsletter (the only place she writes regularly): (your URL)
- Consultation and advisory work: (your booking link)
- Speaking engagements: (your contact email)

CHAPTER

Index

www.ingramcontent.com/pod-product-compliance
Lightning Source LLC
LaVergne TN
LVHW041249110826
845146LV00005BA/1325

* 9 7 9 8 9 9 6 2 6 6 2 6 5 *